Radio and Electronics Cookbook

Edited by
Dr George Brown, CEng, FIEE, M5ACN

Newnes

OXFORD   AUCKLAND   BOSTON   JOHANNESBURG   MELBOURNE   NEW DELHI

Newnes
An imprint of Butterworth-Heinemann
Linacre House, Jordan Hill, Oxford OX2 8DP
225 Wildwood Avenue, Woburn, MA 01801-2041
A division of Reed Educational and Professional Publishing Ltd

A member of the Reed Elsevier plc group

First published 2001

British Library Cataloguing in Publication Data
A catalogue record for this book is available from the British Library

ISBN 0 7506 5214 4

RSGB
Lambda House
Cranborne Road
Potters Bar
Herts
EN6 3JE

Composition by Genesis Typesetting, Laser Quay, Rochester, Kent
Printed and bound in Great Britain

FOR EVERY TITLE THAT WE PUBLISH, BUTTERWORTH-HEINEMANN
WILL PAY FOR BTCV TO PLANT AND CARE FOR A TREE.

Contents

Preface

Although we are surrounded by sophisticated computerised gadgets these days, there is still a fascination in putting together a few resistors, capacitors and the odd transistor to make a simple electronic circuit. It is really surprising how a handful of components can perform a useful function, and the satisfaction of having built it yourself is incalcuable.

This book aims to provide a wide variety of radio and electronic projects, from something that will take a few minutes to a more ambitious weekend's worth. Various construction techniques are described, the simplest requiring no more than a small screwdriver, the most complex involving printed circuit boards.

Originally published by the Radio Society of Great Britain, the projects were all chosen to be useful and straightforward, with the emphasis on practicality. In most cases the workings of the circuit are described, and the projects are backed up by small tutorials on the components and concepts employed. In the 21st century it may seem strange that few of the published circuits use integrated circuits (chips). This is intentional as it is much easier to understand how the circuit works when using discrete components.

Anyone buying the *Radio and Electronics Cookbook* will find that it will lead to hours of enjoyment, some very useful and entertaining gadgets, and increased knowledge of how and why electronics circuits work, and a great sense of satisfaction. Beware, electronic construction is addictive!

WARNING: This book contains construction details of transmitters. It is illegal to operate a transmitter without the appropriate licence. Information on how to obtain an Amateur Radio Licence can be obtained from the Radiocommunications Agency, tel. 020 7211 0160.

1 A medium-wave receiver

Introduction

Let us start off with something that is really quite simple and yet is capable of producing a sense of real satisfaction when complete – a real medium-wave (MW) radio receiver! It proves that receivers *can* be simple and, at the same time, be useful and enjoyable to make. To minimise the confusion to absolute beginners, *no* circuit diagram is given, only the constructional details. The circuits will come later, when you have become accustomed to the building process. In the true amateur spirit of ingenuity and inventiveness, the circuit is built on a terminal strip, the coil is wound on a toilet roll tube (as amateur MW coils have been for 100 years!), and the receiver is mounted on a piece of wood.

Putting it together

Start by mounting the components on the terminal strip as shown in **Figure 1**, carefully checking the position and value of each one. The three capacitors are all the same, and so present no problem. They (and the resistors) may be connected either way round, unlike the two semi-conductors (see later). The resistors are coded by means of coloured bands. You can refer to Chapter 7 if you have difficulty remembering the colours and their values.

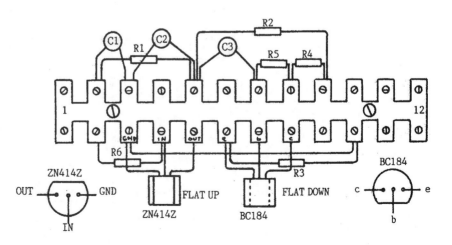

Figure 1 Terminal strip – position of components

1

1. Brown, Black, Yellow   100 000 ohms   (R1, R5, R6)
2. Green, Blue, Brown   560 ohms   (R2)
3. Red, Violet, Brown   270 ohms   (R3)
4. Brown, Black, Orange   10 000 ohms   (R4)

The integrated circuit (the ZN414Z) and the transistor (the BC184) *must* be connected correctly. Check Figure 1 carefully before fitting each device.

Now wind the coil. Most tubes are about 42 mm diameter and 110 mm long. Don't worry if your tube is slightly different; it shouldn't matter. Make two holes, about 3 mm apart, about 40 mm from one end, as shown in **Figure 2**. Loop your enamelled wire into one hole and out of the other, and draw about 100 mm through; loop this 100 mm through again, thus anchoring the wire firmly. Now wind on 80 turns, keeping the wire tight and the turns close together but not overlapping. After your 80th turn, make another two holes and anchor the wire in the same way as before. Again, leave about 100 mm free after anchoring. Using another piece of enamelled wire (with 100 mm ends as before), loop one end through the same two holes which contain the end anchor of the last winding, wind two turns and anchor the end of this short winding using another pair of holes. Figure 2 shows the layout.

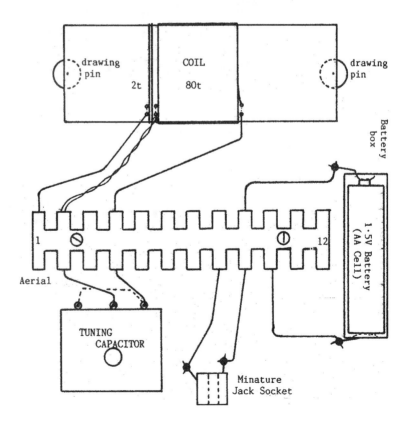

Figure 2 The layout of the parts on the wooden base

With some glass paper, remove the enamel from the ends of both pieces of wire which go through the same holes (i.e. the bottom of the large coil and the top of the small coil), then twist these bare ends together. Remove the enamel from the remaining ends of the coil. The coil is now finished!

The baseboard can be any piece of wood about 150 mm square. Fix the coil near the back edge using drawing pins and connect the wires from the coils to the terminal strip as shown in Figure 2. Using short pieces of PVC-insulated wire (and with assistance if you have never soldered before), solder one piece across the two outer tags of the variable capacitor, shown by the dotted line in Figure 2, and then two longer pieces to the centre tag and one outside tag. Connect these to the terminal strip. Then solder two more insulated wires on to the jack socket (into which you will plug your crystal earpiece), the other ends going to the terminal strip. The last two wires (one must be red) need to be soldered on to the battery box, their other ends going to the terminal strip also. Make sure the red wire goes to the *positive* terminal on the battery, and is connected to terminal 9. The other connection to the battery goes to terminal 10.

Attach the terminal strip to the baseboard with small screws or double-sided sticky tape. The other parts can be mounted the same way.

Listening is done ideally with the recommended crystal earpiece. Don't be tempted to use your Walkman earpieces; they are not the same and will not perform anything like as well. The receiver should work without an extra aerial, but one can be attached to terminal 1 if necessary. A long piece of wire mounted as high as possible is ideal. The *Audio-frequency Amplifier* project will enable you to use a loudspeaker with your receiver, using the signal from the jack socket. No circuit modifications will be needed!

Parts list

Resistors: all 0.25 watt, 5% tolerance

R1, R5, R6	10 kilohms (kΩ)
R2	560 ohms
R3	270 ohms
R4	10 kilohms (kΩ)

Capacitors

C1, C2, C3	100 nanofarads (nF)
	500 picofarads (pF)

Semiconductors
ZN414Z, BC184

Additional items
  12-way 2 A terminal strip
  22 metres of 28 SWG enamelled copper wire
  A few short pieces of coloured PVC-insulated wire
  Crystal earpiece
  3.5 mm jack socket
  1.5 V AA-size battery and box
  Toilet roll tube
  Double-sided sticky tape or selection of screws

Tools required

Small screwdriver, soldering iron.

2 An audio-frequency amplifier

Introduction

This simple amplifier can be built by anyone who is able to solder reasonably well. It doesn't require any setting up and, provided our instructions are followed exactly, will work very well. The circuit diagram is included for the benefit of our more advanced readers, but it is not needed in the construction process. Please practise your soldering before you start, and don't use a printed circuit board (PCB) until you are confident that your soldering is up to scratch.

The amplifier can be used with other projects; it will provide plenty of sound from the MW Radio or from the Morse Sounder projects. It will usually be built into other pieces of equipment, so a box is not supplied with the kit. There is no reason why it shouldn't be put into a box and used as a general-purpose amplifier to help test other projects.

The components

Before you start, you should check that you have *all* the components to hand. A list and some helpful hints are given below.

1. PCB. The plain side is the *component side* and the soldered side is the *track side*. **Figure 1** shows the track side full size. Make the PCB from the pattern given in Figure 1. Otherwise, build the circuit on a matrix board.

2. Three resistors. Locate the gold or silver band around the resistor, and turn the resistor until this band is to the right. There are three coloured bands at the left-hand end of the resistor. Find the resistor whose colours are YELLOW, VIOLET, RED, and look at the resistor colour code chart which you will find in Chapter 7. From this, you will see that YELLOW indicates the value 4, VIOLET the value 7, and RED the value 2. The first two colours represent real numbers, and the last value is the number of zeros (noughts) which go *after* the two numbers. So, the value is 47 with two zeros, i.e. 4700 ohms. In this way, the resistor coloured BROWN, GREY, BROWN has a value of 180 ohms, and the last one, BROWN, RED, GREEN, has a value of 1 200 000 ohms. The ohm (often written as the Greek letter omega (Ω)) is the unit of resistance. If you do not yet feel confident in identifying resistors by their colours, use the Resistor Colour Codes given in Chapter 7.

3. Four capacitors. The two small 'beads' are tantalum capacitors and will be marked 4.7 μF or 4μ7, with a '+' above one lead. A tubular capacitor with wires coming from each end should be marked 220 μF, with one end marked '+' or '−'. This is called an *axial* capacitor because the wires lie on the axis of the cylinder. This is in contrast to the final capacitor, where both wires emerge from the same end. This is a *radial* capacitor, and will be marked 47 μF. Again, one lead will be marked '+' or '−'. Capacitors marked like this are said to be *polarised*, and it is vital that these are placed on the PCB the right way round, so take notice of those signs!

4. Two diodes. These are tiny glass cylinders with a band around one end, and may be marked 1N4148; this is their type number. Like polarised capacitors, they *must* be put on the PCB the correct way round!

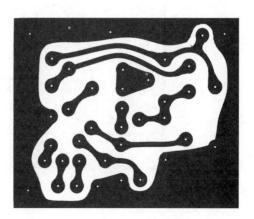

Figure 1 The toil pattern of the PCB – looking from the track side

5. Three transistors. One should be a BC548 (or a BC182), the other two should be BC558 (or BC212).
6. One volume control with internal switch.
7. One loudspeaker. This is quite fragile – don't let anything press against the cone.
8. One PP3 battery clip with red and black leads.

Putting it together

Lay the PCB on a flat, clean surface with the track side downwards. It is always useful to compare the layout with the circuit diagram, given here in **Figure 3**. Although you can't see it, the D-i-Y Radio sign should be at the top. Compare the hole positions with those shown in **Figure 2**. Bend the resistor wires at right angles to their bodies so that they fit cleanly into the holes in the PCB. Push each resistor towards the board so that it lies flat on the board. Then supporting each one, turn the board over and splay out the wires *just* enough to prevent the resistor falling out. Then, solder each wire to its pad on the PCB, and cut off the excess wire. When you have more confidence, you can cut of the excess wire *before* soldering; it often makes a tidier joint.

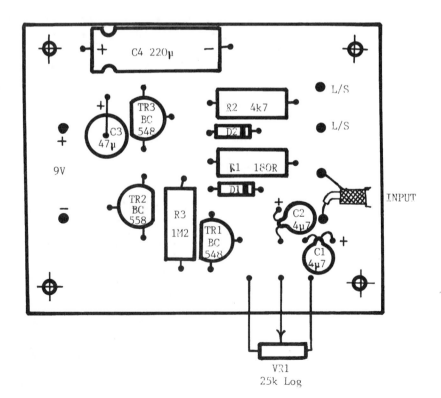

Figure 2 Positions of the components on the printed circuit board (PCB)

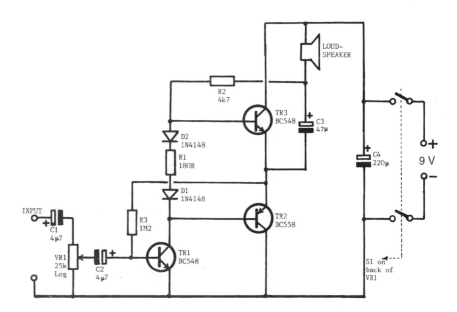

Figure 3 The amplifier's circuit diagram

Now fit the four capacitors. Each must be connected the right way round, so look at each component, match it up with the diagram of Figure 2, bend its wires carefully and repeat the soldering process you performed with the resistors, making sure that the components are close to the board and not up on stilts! Fit the two diodes the correct way round, and solder then as quickly as you can – they don't like to be fried!

Mount the transistors about 5 mm above the PCB. Make sure the correct transistors are in the correct places, and that the flats on the bodies match up with those shown in Figure 2.

Mount the volume control so that the spindle comes out from the front of the board. Use a piece of red insulated wire to the pad marked + on the PCB, and a black piece to the pad marked –, and solder these to the tags on the back of the control, as shown in **Figure 4**. Connect the two leads from the battery clip to the other tags on the switch; Figure 4 will help you. Finally, use two pieces of insulated wire about 100 mm long, twisted together, to connect the loudspeaker to the PCB.

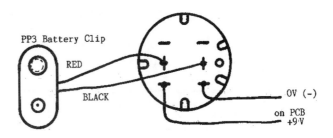

Figure 4 Connections to switch on back of VR1

Box clever!

If you wish to put the amplifier into a box, there is no problem; almost any box that is big enough will do. All that is needed is one hole big enough to accept the bush of the volume control; the PCB will be supported by the volume control. The prototype was not fitted into a box, but mounted on an odd piece of aluminium, bent into an L-shape and screwed on to a wooden base. The loudspeaker was mounted on the aluminium panel by two small pieces of aluminium with 3 mm holes drilled in them, which acted as clips around the edge of the speaker. Drill a few holes in the panel in the position of the speaker to let the sound get out!

Your input signal can be connected to the amplifier with two short pieces of wire, but if the connection needs to be long, use screened cable, with the braid connected as shown in Figure 2.

If you decide to use a different loudspeaker, make sure that its impedance (the resistance value marked on the back of the magnet) is at least 35 ohms. Anything lower may damage TR2 and TR3, and will certainly run down your battery very quickly. You will be surprised at the uses you can find for this little amplifier!

Parts list

Resistors: all 0.25 watt, 5% tolerance
R1	180 ohms (Ω)
R2	4.7 kilohms (kΩ)
R3	1.2 megohms (MΩ)
VR1	25 kilohms (kΩ) log with DPST switch

Capacitors: all rated at 25 V minimum
C1, C2	4.7 microfarads (μF)
C3	47 microfarads (μF)
C4	220 microfarads (μF)

Semiconductors
TR1, TR3	BC548 npn
TR2	BC558 pnp
D1, D2	1N4148

Additional items
PCB
Speaker >35 ohms
PP3 battery clip and battery

3 A medium-wave receiver using a ferrite-rod aerial

Introduction

This design came from the Norfolk Amateur Radio Club, and enables you to build a simple Amplitude Modulation (AM) receiver for frequencies between 600 kHz and 1600 kHz. It should take you around 2 hours to build, and can be used with Walkman-type earpieces. **Figure 1** shows the circuit diagram.

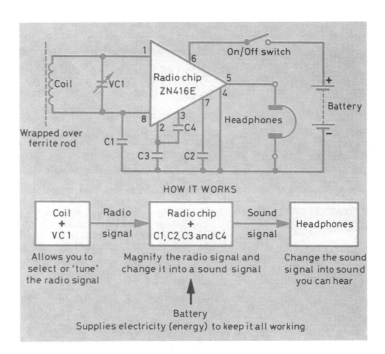

Figure 1 Circuit and block diagrams of the radio

Description

The whole circuit is built on a 50 mm by 50 mm printed circuit board (PCB) designed to fit on the inside of the lid of a plastic box, and is stuck there using sticky pads, the shaft of the variable capacitor going through a hole in

9

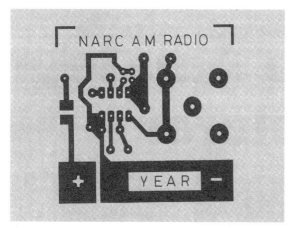

Figure 2a The PCB, solder side

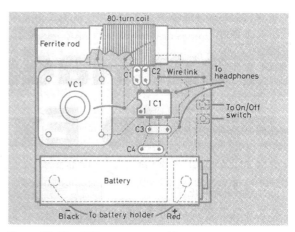

Figure 2b The PCB, component side

the lid. Only two pairs of leads are soldered to the board – one pair goes to the 1.5 V battery in its holder, and the other to the earphone socket. **Figures 2a** and **2b** show the printed circuit and the component layout double size for clarity. You are not obliged to build the circuit on a PCB.

Building it

1. **Check and identify components.** Tick the parts list.
2. **Carefully unwind the wire.** Use paper to make an insulating tube (called a 'former') around the centre of the ferrite rod and secure it with Sellotape. Now, close-wind *all* the wire (leave no gaps between adjacent turns) around the paper former. Secure the winding with more Sellotape, leaving 50 mm of wire free at each end for connection to the circuit. See **Figure 3a**.
3. **Solder in VC1.**
4. **Solder in the integrated circuit holder.** There is a notch in one end of the holder; this should face VC1. Solder also the wire link and the capacitors. Be careful to avoid solder 'bridges' between adjacent tracks on the PCB.
5. **Solder the battery leads.** These must be connected properly – the red battery lead to the + (positive) area and the black lead to the – (negative) area.
6. **Strip bare 1 cm of insulation from the ends of two wires.** Solder them between the PCB and the headphone socket (see **Figure 3b**). Use the end tabs on the socket. Using another pair of insulated wires connect the ON/OFF switch to the PCB tabs shown in Figure 2b.

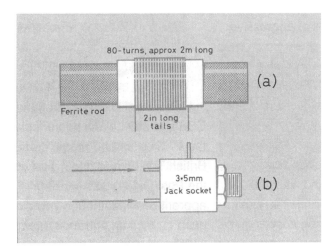

80-turns, approx 2m long

Ferrite rod

2in long tails

(a)

3·5mm Jack socket

(b)

Figure 3 Details of coil and headphone socket

7. **Fix the elastic band.** This goes through the holes at the top of the PCB, with the ferrite rod being slipped through the two end loops. (Note: although the coating on the copper wire is designed to melt away during soldering, it is quite common for difficulty to be experienced in obtaining a good soldered joint; to be on the safe side, remove the coating *before* soldering (with a small piece of sandpaper).) Carefully place the wire ends of the coil through the PCB just above VC1, and solder on the track side.

8. **Fit IC1 into its holder.** This should be done carefully, making sure that *all* the pins are located above their respective clips *before* applying any pressure! Make sure also that the notch on the IC (as shown in Figure 2b) matches the notch in the holder, and faces VC1.

9. **Put battery in its holder.** Listen for some noise in the headphones as VC1 is rotated. Make sure the headphone plug is fully inserted into its socket.

10. **Fix the working board to the lid.** Use the sticky pads and apply *gentle* pressure. Fit the tuning knob, the ON/OFF switch and the earphone socket.

11. **Test again.** If all is still working, fit the lid screws and admire your completed radio!

Parts list

Capacitors
C1, C2	0.01 microfarad (µF)
C3, C4	0.1 microfarad (µF)
VC1	500 picofarads (pF)

Semiconductor
IC1	ZN416E

Additional items
Plastic box (recommended size 76 × 64 × 50 mm internal)
8-pin DIL socket for IC1
Printed circuit board
Tuning knob for VC1
Wire link for PCB
2 m of 30 SWG copper wire, self-fluxing
Piece of paper 25 × 50 mm, to make the coil former
Ferrite rod 70 mm long by 10 mm diameter, approximately
Battery, AA size 1.5 V, with holder and attached wires
Miniature earphone socket (3.5 mm stereo jack)
ON/OFF switch (push-button SPST latched or slide switch)
4 off 100 mm insulated connecting wires, for jack socket and
    ON/OFF switch
Pair Walkman-type earphones
Elastic band, to attach ferrite rod to PCB
4 off sticky pads for securing PCB to box lid

Kits

Ready-made PCBs may be available from Alan J. Wright, G0KRU, Hewett School, Cecil Road, Norwich NR1 2PL.

4 A simple electronic organ

Introduction

This project has nothing to do with radio but, let's admit it, *any* electronics project is good experience! Why not build this little organ – it will keep the children amused at least! It uses the popular NE555 integrated circuit, which contains a circuit which will periodically switch the voltage on the output pin between the supply voltage and zero. Just how frequently this switching occurs depends upon the components external to the integrated circuit. If this switching occurs several hundred or thousand times a second, the change in voltage produced will generate a musical note when connected to a small loudspeaker. The circuit is shown in **Figure 1**.

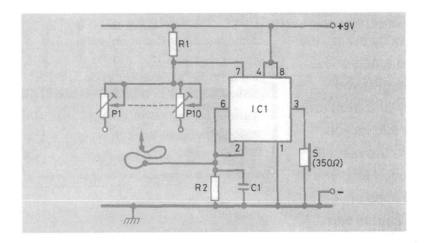

Figure 1 Circuit diagram

Putting it together

(a) Using a PCB. The job is very simple. The placement of components on the unsoldered side of the board is shown in **Figure 2** and the design on the copper track is illustrated in **Figure 3**. Put each component, in turn, on the board, making sure that it lies flat on the board with its tags or wires going cleanly through the holes provided for it; then, solder the wires to the board, cropping them before or after the soldering,

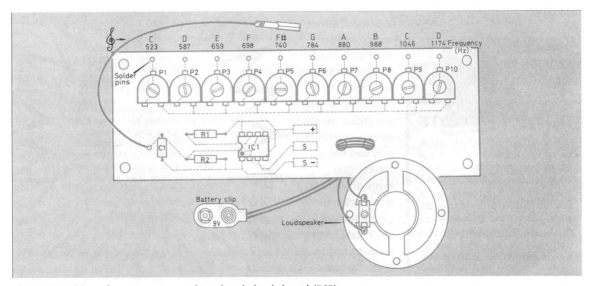

Figure 2 Position of components on the printed circuit board (PCB)

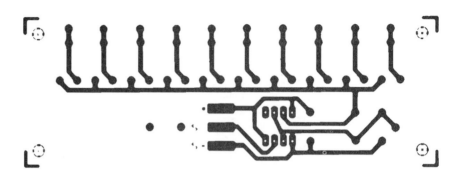

Figure 3 The connections

depending on your preference. If you choose to use a holder for your integrated circuit (highly recommended if your soldering is less than perfect), make sure that the end with a notch in it faces R1 and R2, as shown in Figure 2. Solder the two leads to the speaker to the tabs marked S (either way round), having looped them through the two holes to the right of the tabs in Figure 2. Looping them through the holes acts as a strain relief, ensuring that the soldered joints are not subjected to pulling and bending as you move the wires about. Do the same with the battery leads, the red lead going to the + tab and the negative lead to the – tab (which also has one speaker lead already attached to it). **Figure 4** shows this in detail. Treat the loudspeaker with care – the cone is quite fragile and must not be touched.

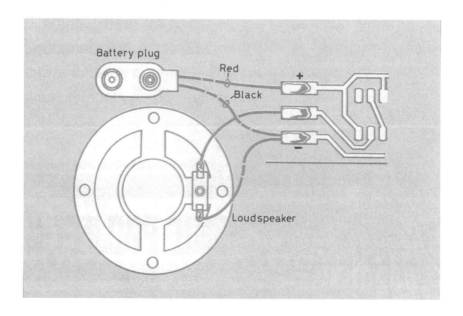

Figure 4 Battery plug and loudspeaker connections

(b) Without a PCB. This is more difficult, and you may need to enlist some help. Using some matrix board (such as Veroboard) is probably the best way of replacing the PCB. You could arrange your circuit in exactly the same way as in the PCB in Figure 3, using wires to replace the copper track.

A simple 'keyboard'

The keyboard is a row of solder pins along the rear edge of the PCB, one for each note covering the range shown in Figure 2. A flying lead with a small spade on it is provided to touch any of the pins in turn, producing any one of ten different notes.

Testing

Check first that each component is in the correct place. When inserting the NE555 chip, first make sure that the end carrying the notch lies over the end of the holder with the notch; then, make sure each pin of the chip lies directly above the hole into which it fits, before pressing *gently* to insert the chip into the socket. Make sure the battery connections are correct, and insert the battery into the clip. Nothing should happen, except for a click from the loudspeaker; touching the spade on any of the pins should produce a coarse note from the speaker. If nothing happens, check everything again; don't *assume* that wires go where you think they go!

After you get the first note, all the others should work, too, but they will sound off-tune at first. The organ needs tuning up by adjusting the 10 preset variable resistors P1 to P10. The approximate frequency to which each note should be tuned is given in Figure 2; if you can beg, borrow or steal a frequency counter, setting up is easy. If you have a piano, the organ can be tuned by comparison of the notes with those on the piano. The frequencies are given in Hertz (abbreviation Hz), and represent the number of times the IC switches on and off every second. If the sound coming from the loudspeaker is too loud or very distorted, then try putting an $330\,\Omega$ resistor (colour code orange, orange, brown) in series with the loudspeaker. This is done by taking the resistor and cutting its leads to about 5 mm; then, disconnect one speaker lead from the tab on the PCB (it doesn't matter which). Solder one end of the resistor to the vacated speaker tab, and the free speaker lead to the other end of the resistor. This will limit the volume of sound from the speaker, and lengthen the life of your battery. If it is still too loud, try a resistor of a larger value, or use a smaller resistor to make it louder.

Parts list

Resistors: all 0.25 watt, 5% tolerance
   R1   2.7 kilohms (kΩ)
   R2   1 Megohm (MΩ)
   P1–P5   Preset resistor 100 kilohms (kΩ)
   P6, P7   Preset resistor 50 kilohms (kΩ)
   P8   Preset resistor 25 kilohms (kΩ)
   P9, P10   Preset resistor 10 kilohms (kΩ)

Capacitor
   C1   100 nanofarads (nF) or 0.1 microfarad (μF)

Integrated circuit
   IC1   NE555 timer chip

Additional items
S   Loudspeaker >60 ohms (Maplin)
               1 off battery clip (for PP3 battery)
               1 off spade terminal
               12 off solder pins 'Veropins'
               3 off 10 cm lengths of 'hook-up' wire

(This article is based on projects originally designed by Radio Scouting, Netherlands.)

5 Experiments with the NE555 timer

Introduction

Several of the projects in this book use the NE555 timer, an integrated circuit which is at the heart of many circuits whose processes are determined by time intervals. **Figure 1** shows the circuit diagram of an audio oscillator using the 555. The timing voltages (governing the frequency of oscillation) are produced by R1, R2 and C2; a voltage appears at pin 3 which 'switches' at this frequency between zero and a voltage close to the supply voltage, which in this case can be anywhere between 6 V and 14 V. The output current, when applied through R3 to a small loudspeaker, produces an audible tone, provided that there is a DC path between the two test leads.

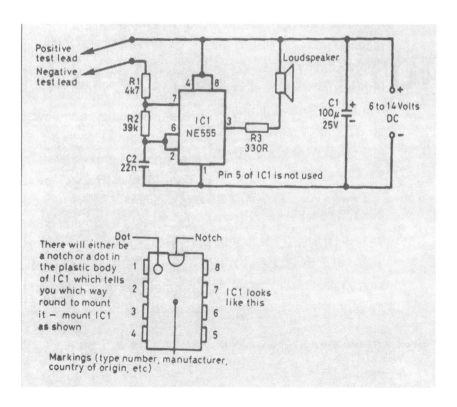

Figure 1 The circuit diagram

Construction

The simplest way to mount the components is on a piece of matrix board (Veroboard), available from any of the good suppliers. The prototype of this circuit used the type of board with copper strips along the underside; these strips are used like the copper tracks on a PCB, to join components together. Firstly, cut the four strips between the positions of the pins of the IC socket, as shown in **Figure 2**. You can buy a tool for this purpose, but a small twist drill (about 3 mm diameter) is just as good. Turn it between your fingers – if you use a drill you will end up with holes right through the board! Then solder in the IC socket (with the notch in the position shown), followed by the four links made with single-conductor insulated wire. Put in each component as shown, ensuring that C1 (an *electrolytic* or *polarised* capacitor) is connected correctly. When all the components have been soldered in, take the 555 chip and lay it on its socket, with its own notch lying above that of the holder. Then, making sure that each pin lies *directly* above its corresponding socket, press down *gently* on the chip, with the board supported on a flat surface.

Testing

Connect the circuit to a battery or small power supply, ensuring that the positive and negative leads are the right way round. Always use red and black leads here, then you are less likely to get it wrong! Switch on. Nothing

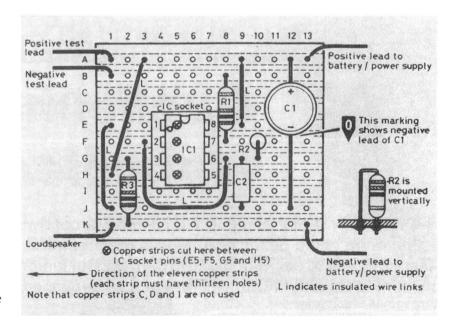

Figure 2 Veroboard layout. If you can read a circuit diagram, the project can be built using other methods

should happen until you short together the two test leads, when there should be a note from the loudspeaker. If this doesn't happen, switch off, disconnect the circuit and check your wiring and soldering. Is it exactly like Figure 2? Are the soldered joints round and shiny? If any are dull, then 'sweat' them briefly with a hot soldering iron until the solder runs, remove the iron, and check that they are as shiny as the rest. Check that there are no solder 'bridges' between adjacent tracks by holding your board up to a strong light. Then, reconnect, switch on and touch the test leads together. All should now work!

Uses of your circuit

1. As a Morse practice oscillator. Simply connect the two test leads to your key and, each time the key is pressed, you should hear a note from the speaker. The frequency of the note may be altered by putting a resistor in series with the key. To do this, remove one test lead from the key and select a resistor; connect one end of the resistor to the free test lead and the other to the empty terminal on your key. Selecting the value of resistor that you need will be a useful experiment in itself.

2. As a continuity tester. You can check fuses and lamp bulbs by connecting them across your test leads. If the speaker remains silent, the fuse or bulb has blown.

3. To indicate changes of resistance. Hold the ends of the test leads in each hand; you should hear a low note, because of the high resistance of your body. Squeeze the ends harder, and the frequency of the note should rise, because you are now making better contact. Repeat this with damp hands and the frequencies will be higher still.

4. As a thermometer. Connect the test leads to a thermistor (a device whose resistance changes with temperature) and warm it with a hair-dryer, or even in your hands, and you will hear the pitch changing with the temperature of the thermistor. A suitable 'bead' thermistor is available from Maplin (order code FX21).

5. As a diode tester. Use any diode, and connect the negative test lead to the end of the diode marked with the ring. This is the cathode of the diode. The other end, the anode, should be connected to the positive test lead, and a note should be heard from the speaker. This does not necessarily mean that the diode is working – yet. Reverse the connections and *nothing* should be heard. If this is the case, the diode *is* working.

6. As a light meter. Use a photoconductive cell (a device whose resistance changes with light intensity) connected between the test leads. A note should be heard. Shading the device with your hand will increase its resistance and the note should decrease in frequency. A suitable device is the ORP12 cell from Maplin (order code HB10).

There are many more applications. **Do not connect the test leads to other circuits that are switched on.** Your circuit, or the circuit you are connecting it to could be damaged. Think of a *passive* device or circuit (i.e. one *not* requiring a power supply or battery) where changes of resistance occur, and you have found another application!

Parts list

Resistors: all 0.25 watt carbon film types
    R1   4.7 kilohms (kΩ) – yellow, violet, red
    R2   39 kilohms (kΩ) – orange, white, orange
    R3   330 ohms (Ω) – orange, orange, brown

Capacitors
    C1   100 microfarads (μF) 25 V radial electrolytic
    C2   22 nanofarads (nF) or 0.022 microfarad (μF) polyester
            type with 10 mm lead spacing

Integrated circuit
    IC1   NE555

Additional items
    Miniature loudspeaker (preferably 35, 40 or 80 ohm)
    8-pin DIL socket for IC1
    0.1 inch Veroboard ('stripboard'), size 11 strips by 13 holes
    PVC-covered stranded wire for test leads, loudspeaker and battery
        connections
    PVC-covered solid wire for links on the board
    A power source of between 6 V and 14 V, such as a 9 V battery (PP3)

Component sources

Cirkit
Tandy – many high street shops

6 A simple metronome

Introduction

A metronome is a device used by musicians to indicate the tempo of a piece of music. Until electronics came on the scene, this 'beating of time' was achieved in much the same way as a clock keeps time, i.e. with a pendulum device, the clicking of the escapement indicating the beats of the music.

Those of you who have already built the Morse Key and Buzzer from the designs in this book, will recognise the circuit of this metronome – it is exactly the same as was used to produce the note of the buzzer. This circuit is shown in **Figure 1**.

The circuit

Three components determine the speed at which the circuit oscillates – the speaker (LS), the resistors (VR1 + R1) and the capacitor (C1). VR1 is a variable resistor, so that the speed at which the oscillator operates can be varied. Compared with the component values of the Morse Buzzer (which operated at around 800 Hz), these components now give an oscillation frequency of around 1.25 Hz, which is far too low to be heard as a note. What we *do* hear, however, is a series of clicks, as the voltage across the speaker changes quickly from 0 to 9 V and back again.

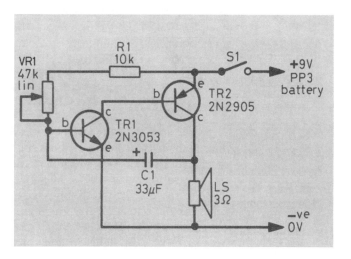

Figure 1 The metronome circuit is rather like the Morse oscillator

21

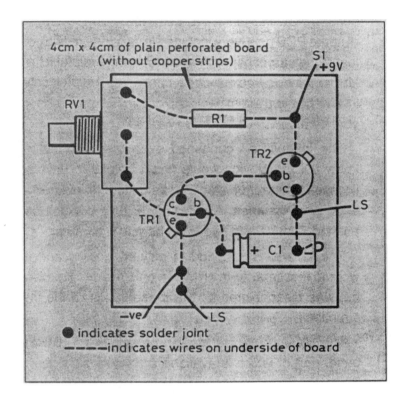

Figure 2 The component wires are pushed through holes in the circuit board and joined together underneath

Variation of speed could be achieved by varying resistance or capacitance. However, as you may already know, variable capacitors have values in the picofarad range, not the tens of microfarads used here, so it is very simple to employ a variable resistor (potentiometer) to control the oscillator. You could use a multi-way switch to switch in one of several capacitors, as well as having the variable resistor, but this was found to be an unnecessary complication. This design operates between about 100 clicks per minute and 200 clicks per minute.

Making the prototype

A single piece of plain matrix board (no copper strips) measuring about 40 × 40 mm is sufficient to hold all the components except the potentiometer and switch (see later). The case can be plastic or aluminium, and one measuring 65 × 100 × 50 mm is about right. Make sure there are holes in the case beside the speaker cone to let the sound out, and larger holes for the potentiometer and switch. If a potentiometer is used with a combined ON/OFF switch, then the extra hole for the switch is not necessary! It is advisable to construct the circuit *before* putting it in the box, so that it can be tested to ensure that everything is working. If it is, then you can exercise your ingenuity in mounting the speaker, battery and board inside the box. A final test can be made before starting the calibration process.

Calibration

There is no 'easy' way to do this. The frequencies involved are too low to be measured with the average frequency counter, so you will need to resort to using a stopwatch and counting the number of clicks per minute.

Parts list

Resistors: 0.25 watt, 5% tolerance
R1	10 kilohms (kΩ)
VR1	47 kilohms (kΩ) linear potentiometer

Capacitor
C1	33 microfarads (μF) electrolytic

Transistors
TR1	2N3053	npn
TR2	2N2905	pnp

Additional items
S1	SPST	ON/OFF switch
LS	3 ohms (Ω)	loudspeaker
	Knob with pointer for VR1	
	PP3 battery and connector	
	Aluminium case, 65 × 100 × 50 mm	
	Matrix board (plain), 40 × 40 mm	

7 What is a resistor?

Introduction

Materials that carry electricity easily are called **conductors**. They include all metals and salt water, for example. We use wire as a conductor, and the ease with which it passes an electric current depends upon the material, its thickness and its length. Silver (symbol Ag), gold (Au), copper (Cu) and aluminium (Al) are the best metallic conductors. Most wires are made of copper, although the best conductor, weight for weight, is aluminium.

Materials that *don't* carry current (or, at least, do so very badly) are called **insulators**, and they include dry wood, rubber, plastic and glass among their number. Wires are often coated with a layer of insulator to prevent adjacent wires touching and causing an accident.

Resistors

If there wasn't such a thing as resistance, the subject of electronics wouldn't exist; only infinite currents would flow and voltages wouldn't exist either! We need to reduce the flow of current if we are to make current do something useful for us. Components that resist the flow of current are called **resistors**, and they are said to have a **resistance** which is measured in ohms (Ω), named after Georg Ohm, who formulated the law (also named after him) by which the voltage and current through a conductor are related. His law gave rise to the formula *everyone* remembers:

$$I = \frac{V}{R},$$

where I is the current flowing, measured in amps,
V is the voltage across the conductor, and
R is the resistance of the conductor, measured in ohms.

From this equation, you can see that, for a constant value of voltage, V, if the resistance goes up, the current will go down, and vice versa. The circuit symbols for resistors are shown in **Figure 1**. You will find the upper symbol in older magazines; it is still preferred by many engineers. The lower symbol is the prevalent standard symbol.

Resistors are made in several ways, the cheapest using carbon; another type is usually made from a ceramic cylinder (used only as a support) on which

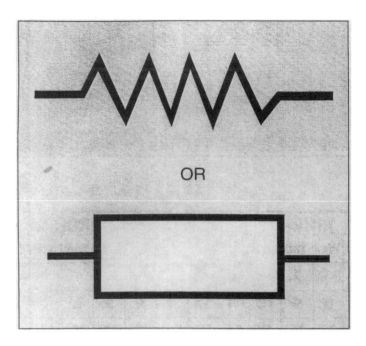

Figure 1 Circuit symbols for resistors

is placed a very thin film of metal – the thinner the film, the greater the resistance. All resistors are coated with a thin film of insulation, for the same reason we discussed earlier.

The colour code

Each resistor has coloured bands on it which enable us to see what value of resistance it has. There are normally three (but sometimes four) at one end, and a single one at the other (see **Figure 2**). The colours indicate figures, according to the list below.

Colour	Value	Colour	Value
Black	0	Green	5
Brown	1	Blue	6
Red	2	Violet	7
Orange	3	Grey	8
Yellow	4	White	9

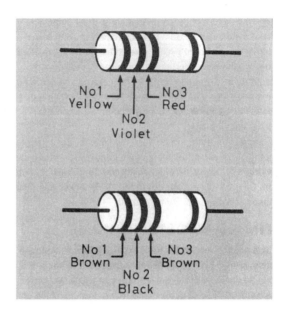

Figure 2 Some examples of resistor colour codes; top 4700 Ω (4.7 kΩ) and bottom 100 Ω

Using the colour codes is easy, once you see the logic behind it. Hold the resistor so that the single band is towards the right. The three colours on the left are read in the normal order from left to right. The first two bands *always* indicate numbers; the third band gives the number of zeros to add to the right of these two numbers. So, looking at the top resistor in Figure 2, yellow, violet, red means 4, 7, and two zeros, giving 4700 ohms! Looking at the lower resistor, brown, black, brown means 1, 0, and one zero, giving 100 ohms.

Remembering the order of the colours may be difficult at first. The colours from red to violet are the colours of the rainbow, in order, so if you know those, you're almost there! Around those colours are black and brown below the red, and grey and white above the violet, which you can imagine as getting brighter from black to white (well, almost!). It won't be long before you don't need to remember them at all.

The isolated band on the right-hand side is not part of the resistor's value; it indicates its *tolerance*, i.e. how close it might be to the indicated value. A brown band indicates ±1%, a red band ±2%, a gold band ±5% and a silver band ±10%. For example, a resistor marked as being 100 ohms with a ±5% tolerance will have an *actual* value somewhere between 95 ohms and 105 ohms.

8 Waves – Part 1

Introduction

Waves are responsible for most of the processes in life where energy is transferred from one place to another. Heat and light energy from the sun, for example, come to us as *electromagnetic waves*. Sound travels through the air as a wave; it is not the same sort of wave as light or heat, but it obeys many of the same properties. Damage is caused to coastal margins by the waves of the sea – again, another type of wave, but still obeying many of the same properties. Radio waves are of the same type as heat and light waves, as are gamma rays, X-rays, ultra-violet waves and infra-red waves. So, once we begin to understand what radio waves do, we are also learning about a huge chunk of physics at the same time! All these waves are part of the *electromagnetic spectrum*. The word 'spectrum' simply means a 'range', so what we have is a range of electromagnetic waves – that's all!

Sensing things

Light waves are invisible, but our eyes can detect the *effect* they have on different materials because the waves produce an effect on the retina of the human eye which the brain can interpret. We cannot see heat waves either, but we can feel the *effect* they have on our skin. Gamma rays and X-rays are also invisible, but their detrimental *effects* on human tissue are well known. It is not surprising, then, that we cannot see radio waves. We cannot sense them, either, until we produce a device upon which they have an *effect*. That device is a **radio receiver**; it is able to process certain characteristics of radio waves, and make these characteristics audible by generating sound waves from the loudspeaker or headphones. Other characteristics of the same waves may be turned into light as a TV picture on a cathode-ray tube, or as a fax image on a sheet of paper.

Visible waves

Let's start our description with some waves that we can actually see! When a small stone is thrown into a pond, we see circular water waves *radiating* from the point where the stone fell into the water, as **Figure 1** shows. (Notice that we use the word *radiating*, even with water waves; it is not a radio term, but one which describes any motion where the *radius* of a circle is increasing. In this case, it is the radius of the circular waves which is

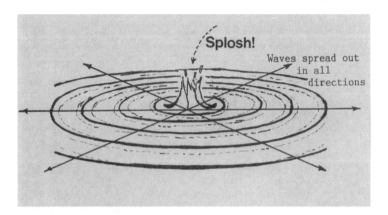

Figure 1 The waves move out from the point where the stone landed

increasing.) If you were to watch the water waves down at the water level, perpendicular to the direction in which they are travelling, you would see something like the illustration in **Figure 2**.

The horizontal line represents the water level before the wave started, and the vertical line represents the direction in which the water is displaced at any instant.

All waves are described in the same way

'Freezing' the motion of the water in this way allows us to define two very important characteristics of a wave, characteristics which we talk about every day – *wavelength* and *amplitude*. The wavelength of a wave is simply the distance (measured in metres) from any point on one wave to the same point on the adjacent wave. Look at the diagram and you will see what is meant by 'the same point on the adjacent wave'. The amplitude of a wave

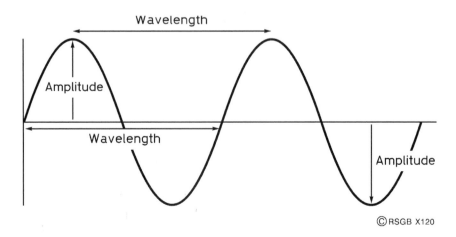

Figure 2 A water wave, viewed in cross-section

©RSGB X120

is always measured from the centre (undisturbed) position of a wave to the peak (or the trough) of the wave. Both these positions are shown, the arrow indicating that the measurement is taken *from* the centre *to* the peak or trough. The amplitude of a wave is defined as the maximum displacement of the wave from the centre position – the direction (up or down) of that displacement does not matter. Waves of greater amplitude carry more energy with them.

If we now 'unfreeze' the wave, we will see it travel from left to right (or right to left, depending on where we are looking). The speed at which it moves is called its *velocity*, and is measured in metres per second.

Another useful word is *propagate*; it means *travel*. We talk about radio waves propagating from a transmitter to a receiver. This velocity of propagation (for electromagnetic waves) is very fast indeed – they will cover 300 million metres in one second. This is virtually incomprehensible, so think of a radio wave travelling around the earth – it can travel $7\frac{1}{2}$ times round the earth in one second! We use the symbol c for the velocity of radio waves (which is the same as the velocity of light, of course – all electromagnetic waves travel at this speed through air and space).

The last thing we need to know about the wave is its *frequency*. Imagine a cork floating on the water in the path of the wave; it will bob up and down. If we were able to count the number of times it went through its highest position in one second, then that number would be its frequency. Any periodic motion like this is said to go through one *cycle* each time one complete wave passes a point (in this case, our cork). We are thus counting the number of cycles per second of the cork's motion. The unit of frequency is thus 'cycles per second'; this unit is now named after Hertz, a radio pioneer, and is abbreviated to Hz.

Our description of the wave is now quite simple – we need only four quantities:

(a) **Frequency**   symbol f – unit, hertz (Hz)
(b) **Wavelength**   symbol λ (Greek letter lambda, pronounced 'lamb-da') – unit, metre (m)
(c) **Amplitude**   symbol a – unit depends on application
(d) **Velocity**   symbol c – unit, metres per second (m/s or ms^{-1})

The basic formula

Whatever may happen to a wave while it travels through different media (vacuum, air, brick, wood, etc.), one thing and **only** one thing remains constant – its frequency. Its wavelength, amplitude *and* velocity may change, but its frequency never does. Three of the four characteristics already identified are connected by the simple relationship

$c = f \times \lambda.$

Remember that c is constant if the wave travels in air or in a vacuum. This means that waves having higher frequencies (f large) must have smaller wavelengths (λ small) and vice versa. You can imagine frequency and wavelengths being on opposite ends of a see-saw!

Divisions of units

Because the frequencies of radio waves are so high (despite them having the lowest frequencies in the electromagnetic spectrum!) we have a problem with writing them down. Do you write in your log book that you have just heard a station on 14 100 000 Hz? Of course not, you write it as 14.1 MHz, knowing that the prefix mega (M) means 'one million'. The prefixes which you need to know (when applied to frequency) are:

kHz	kilohertz	meaning	1000 Hz
MHz	megahertz	meaning	1 000 000 Hz
GHz	gigahertz	meaning	1 000 000 000 Hz.

Notice that the 'k' in kilohertz is a lower case letter. It is **incorrect** to write it as an upper case letter. 'K' is a computer-related prefix meaning **not** 1 000 but 1 024!

When we come on to discuss heat and light waves, we will use wavelengths rather than frequencies, because of the see-saw effect – as the frequencies get larger and larger, the wavelengths get smaller, and hence are numbers which are more manageable, both to talk about and to write down!

Bands

Gamma rays, X-rays, ultra-violet waves, light waves, infra-red waves are all part of the electromagnetic spectrum, but we divide them up because they have different properties. This is why we divide up our radio frequencies into different bands. The radio waves of top-band signals (around 2 MHz) have completely different properties compared with those in the 20 metre band, so we are dividing up the *radio spectrum* in the same way – by property.

9 A beat-frequency oscillator

Introduction

Many readers will know that, although they have a short-wave radio which covers at least one of the amateur bands (e.g. 7 MHz or 14 MHz), they are unable to listen to SSB or Morse signals. This is because the receiver lacks a Beat-Frequency Oscillator (BFO). We need the 'carrier' frequency of a BFO to replace the carrier that has been removed from the signal at the transmitter. When listening to Morse signals, the BFO signal 'beats' with the incoming signal to produce a note in the loudspeaker. If you are a musician, you will be familiar with the method of using 'beats' to tune one musical instrument from another; in the BFO, the beat frequency produced is the tone signal you hear.

In the more complex amateur radio receiver, a BFO is incorporated as part of the whole system. In our model, it is an external circuit that sits alongside your radio. The circuit diagram is shown in **Figure 1**.

Construction

Built on a small piece of matrix board about 80×50 mm, the circuit can be fitted inside a small plastic box. For once, we *don't* want to screen the circuit to prevent it interfering with other equipment; we *want* it to interact

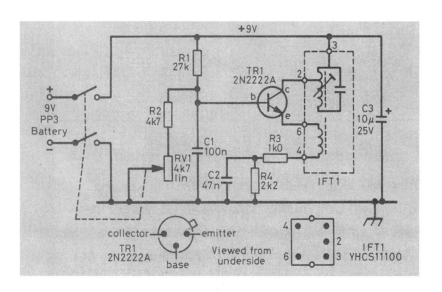

Figure 1 Circuit diagram of the BFO

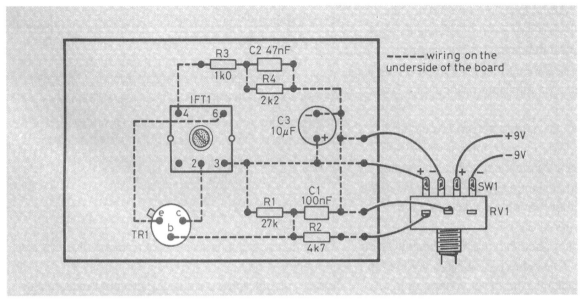

Figure 2 Matrix board layout shown from the component side. Adjust IFT1 carefully for the best results

with our receiver! This is why we use a plastic box. Maplin Electronic Components supply a suitable box, complete with the matrix board to fit inside (order code YU46).

Look at **Figure 2** carefully before you start to build the circuit, so that you can position the components correctly. Firstly, inspect the intermediate frequency (IF) transformer, IFT1, and remove its metal screening can very carefully. Again, this is to allow some signal to escape from our circuit and enter our radio. Having done this, solder the components, using the matrix board as a support. Underneath the board, the components are linked by single-conductor, insulated wire. Take particular care with the polarity of the electrolytic capacitor, C3, and the connections to the transistor, TR1 (see Figure 1).

The variable resistor, VR1, has a switch mounted behind the control itself, and the insulated leads to it from the battery should be about 10 cm long. Connect these before fitting VR1 into the case, so that the BFO can be calibrated (adjusted) correctly.

Calibration

After a final check that all the components have been fitted and soldered correctly, connect the battery, switch on, and hold the transistor between your fingers, to check that it is not getting hot. Place the circuit close to your receiver, and set RV1 to mid-position. Tune your receiver to find an amateur SSB transmission; the frequencies listed below will help you in knowing where to look. It may sound very strange, but don't worry. *Slowly* turn the core of IFT1 with a small, non-metallic screwdriver or with the correct 'trimming tool'. The core into which the blade fits is very fragile, so attempt this process with care. When the speech sounds as natural as you can get it, leave the core at this position, and use RV1 to make the speech sound natural.

Using the BFO

For best results, you may have to move the BFO nearer or further away from your radio. At the lower end of most bands (for instance just above 7.000 MHz or 14.000 MHz) you should be able to resolve Morse code (CW) signals. If you find that the BFO signal is a little weak, solder a 15 cm length of insulated wire to pin 2 of IFT1, and place it alongside your radio. This should improve signal intelligibility. When you are happy with the performance, switch off, drill a 10.5 mm hole in the box and fit RV1, followed by the matrix board assembly. Screw the base to the box, fit the knob, and the BFO is complete!

Where to listen

Band	Frequencies (MHz)
15 m	21.000–21.450
17 m	18.068–18.168
20 m	14.000–14.350
30 m	10.100–10.150
40 m	7.000–7.100
80 m	3.500–3.800

Parts list

Resistors: all 0.25 watt, 5% tolerance

R1	27 kilohms (kΩ)
R2	4.7 kilohms (kΩ)
R3	1 kilohm (kΩ)
R4	2.2 kilohms (kΩ)
VR1	4.7 kilohms (kΩ), linear, with DPST switch

Capacitors

C1	100 nanofarads (nF) or 0.1 microfarad (μF), ceramic
C2	47 nanofarads (nF) or 0.047 microfarad (μF), ceramic
C3	10 microfarads (μF), 25 V radial, electrolytic

Additional items
TR1	2N2222A npn
IFT1	Toko type YHCS11100
Box	plastic, approximate size $100 \times 70 \times 45$ mm
Board	matrix, to fit inside the box
Connector	for PP3 battery
Knob	for RV1

10 What is a capacitor?

Introduction

A capacitor is a device that stores energy in the form of electricity. Much less energy than a battery, and for a shorter time, however. The simplest form of capacitor takes the form of two flat metal sheets separated by air; connections are made to each plate, as **Figure 1** shows. If you imagine a pair of these plates, 30 cm square and separated from each other in air by 1 mm, the capacitance of this device would be almost exactly 80 picofarads (pF), i.e. 80 million-millionths of the unit of capacitance, the farad. Now this is quite a small value, you will agree, and it comes about because the farad is such a large unit. Nevertheless, as you will probably know, we may have capacitors of value 10 000 microfarads (μF) in our radio equipment, and they can be smaller than your little finger, so they are obviously not made the same way!

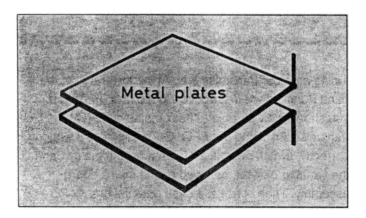

Figure 1

Large and small

We cannot go into the manufacture of capacitors here – after all, we are users of the devices, not the designers! First of all, beware of incorrect statements; the words 'capacitor' and 'capacitance' are *not* the same. For example, a large capacitor would be a description of one the size of a toilet roll. It need not have as large a capacitance as one the size of your little finger. A 'large capacitor' is one which is physically *big*, a 'large capacitance' refers to a capacitor which can store a larger amount of energy when a certain voltage is applied between its plates. The capacitors in a mains power supply are usually big **and** have large capacitances. High-power RF amplifiers may have large capacitors with small capacitances!

Electrolytics . . .

Electrolytic capacitors usually have capacitances of $1\,\mu F$ or above. They differ from other capacitors in that they **must** be connected the right way round (i.e. they have positive and negative connections, just like a battery). They may explode if the connections are reversed!

. . . and the others

Other capacitors may be connected either way round, despite their names. We have *polystyrene*, *ceramic*, *silver-mica* and *tantalum*. Each has its own advantages and disadvantages, and the parts list for a project will always tell you which type is best.

Storing energy

If you were to connect a large capacitance across a 12 V power supply, nothing would appear to happen. Removing the capacitor from the supply and connecting it to a voltmeter would show that the capacitor has 12 V between its ends. This shows that, while the capacitor was connected to the supply, energy flowed from the supply into the capacitor. We say that the capacitor was *charged up* by the supply. If you are using an *analogue* voltmeter (i.e. one with a meter and pointer), you will notice that the indicated voltage slowly drops until, eventually, there is no voltage across the capacitor. This is because the capacitor has *discharged* its energy into the voltmeter. If you had used a smaller capacitance, the same would happen, except that the voltage would drop to zero more quickly – the capacitor stores a smaller amount of energy because its capacitance is smaller. Capacitors behave like other things in life – a small car can move more quickly than a large bus – a small piccolo emits a higher note than a flute – the voltages in a circuit containing a small capacitance will change more quickly than those in a circuit with large capacitance.

Varying the capacitance

Some capacitors are capable of having their capacitance changed manually; these are called *variable capacitors*. They work like the basic capacitor of Figure 1. Imagine moving the top plate of the pair a little to one side; the capacitance is determined, not just by the size of the two plates, but by their *area of overlap*. As this decreases, so does the capacitance. Such devices are limited in their capacitance, about 500 pF being the maximum value.

AC and DC

Because the plates of a capacitor do not touch each other, a direct current (DC) cannot pass between them. However, an alternating voltage on one plate can induce an identical alternating voltage on the opposite plate, and thus a capacitor appears to pass an alternating signal, even though currents as such, do not pass between the plates. This property of passing AC and not DC is very important, and a capacitor used in this way is called a *DC blocking capacitor* or, simply, a *blocking capacitor*. A blocking capacitor can be used at the same time, to couple a signal from one circuit to the next; here it would be known as a *coupling capacitor*. *Decoupling capacitors* are to be found where the capacitor is employed to remove an AC signal while retaining a DC component.

Finally. . .

Unlike resistors, the manufacture of capacitors renders them susceptible to excess voltage, so if you find a capacitor labelled 10 μF 16 V, it means that operating it above 16 V may fatally damage the device (and the circuit around it). This voltage is called the *working voltage* of the capacitor; on some electrolytics, you may find it expressed as *volts working* (i.e. 8 μF 450 V WKG).

Many smaller capacitors have their properties marked on them in a colour code, like resistors. **Figure 2** shows these codes, and their meaning, and the table below summarises the values of the colours.

Table 1

Colour	Value	Voltage (tantalum capacitor)	Voltage (polyester capacitor)
Black	0	10	–
Brown	1	–	100
Red	2	–	250
Orange	3	–	–
Yellow	4	6.3	400
Green	5	16	–
Blue	6	20	–
Violet	7	–	–
Grey	8	25	–
White	9	3	–

Figure 2 Some capacitors have coloured bands or stripes, rather like resistors. The colour code, which is the same as the resistor code, is shown in Table 1. The band shown on the chart as '1st' is the first number of the capacitor's value in pico-Farads, '2nd' is the second number and 'M' is the Multiplier or number of noughts. For example, a capacitor reads from the top: Brown, Black, Yellow, Black, Red. Its value is One, then Nought, then Four more noughts = 100 000 pF (also referred to as 0.1 μF or 100 nF). Its tolerance (Black) is 20% and the working voltage (Red) is 250 V. The 'V' means the maximum working voltage. The band marked 'T' shows the tolerance, just like resistors, and the one marked 'TC' is only used on special capacitors designed to change their value with temperature

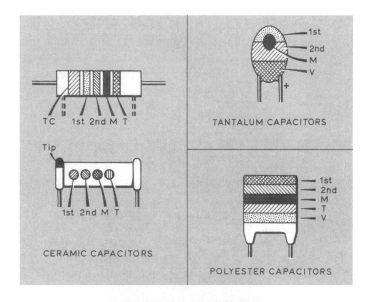

11 Waves – Part 2

Introduction

We left Part 1 with the concept that radio waves are divided up into bands which have different properties. Not *all* the properties are different, though. We need to discuss several wave properties, so we will start with what happens to waves as they propagate over long distances.

Getting weaker

Imagine a torch battery connected to a bulb by wires about 1 metre long. The bulb lights normally. If we now take the bulb 100 metres away from the battery and wire it up, we would expect the bulb to be somewhat dimmer, which is exactly what would happen. It happens because of the resistance of the wires – the wires do not form a perfect conductor. A similar situation occurs with radio waves.

All waves suffer from *attenuation* – they get weaker the further they travel. In cases of extreme attenuation, we need to apply some *amplification* before the attenuated wave can be used in a receiver.

Carrying information

When we speak over the telephone, the range of frequencies in our voices extends from very low frequencies up to about 15 or 20 kHz. In audio terms, this is quite a large *bandwidth* (meaning a wide band of frequencies). For communications purposes, however, most of this bandwidth is not needed, and in the telephone system (and in our transceivers), this is cut down so that it extends from about 200 Hz to 3 kHz, a reduction in bandwidth from 20 kHz to about 3 kHz. A bandwidth of 3 kHz has been found to be sufficient to convey speech intelligibly which, after all, is just what we need!

The radio waves coming from an amateur transmitter convey our speech signals over long distances. By themselves, the speech signals do not travel very far, so they have to be combined with a radio signal that *will* travel long distances. This extra signal is called the *carrier wave* (or just the *carrier*), because its job is to *carry* the speech signals along with it! The process of combining the speech (or Morse code) with the carrier is called *modulation*.

Wider and wider

Sending Morse code is achieved simply by switching (or *keying*) the carrier on and off. The bandwidth of the transmitted signal is only about 100 Hz. Speech, with its reduced bandwidth of 3 kHz, will produce a *single-sideband* (SSB) transmitted signal with a bandwidth of 3 kHz. If the same speech signal were used to produce an *amplitude-modulated* (AM) signal from the transmitter, it would have a bandwidth of about 6 kHz. Perhaps you can now understand why the bandwidths needed to produce hi-fi broadcasts need to be so large. TV signals need bandwidths running into tens of megahertz!

Waves need aerials

Radio waves are produced whenever changing currents flow through a wire, and when that wire is made in such a way as to maximise the *radiation* from the wire, it is called an *aerial* or *antenna*. The same piece of wire will receive radiation from other aerials; an aerial will transmit and receive. This is an important property of the aerial: when a current flows through it, electromagnetic waves are launched into the air; when electromagnetic waves in the air encounter the aerial, currents are produced in it.

From the simplest transmitting aerial, waves travel in all directions, like the waves on the pond that we considered in Part 1. They will travel a long way through air and space before they become too week to be received. They won't travel very far into the earth, however! The earth will *reflect* some of the wave and will *absorb* the rest. That portion of the wave which is reflected will again travel through air and space until it is totally attenuated.

Look at **Figure 1**; A represents a radio transmitter, with B1, B2 and B3 being receiving stations. The two arrows pointing 'downwards' from A represent two of the waves from A which just graze the earth's surface. Waves above

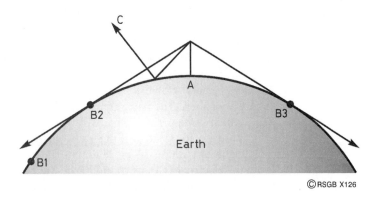

Figure 1 Stations between B2 and B3 receive the *ground wave*

©RSGB X126

these will travel on into space; waves below them will either be absorbed and reflected by the earth or received by aerials. B1 will not receive any signals from A, because it is below A's horizon. B2 and B3 can receive A's signals because they are *just* on A's horizon. Any stations between B2 and B3 will also receive A's signals, which are known as *ground-wave signals*. The wave at C represents one which is reflected by the ground and travels into space.

This description begs the question of how signals are received from stations well beyond the ground-wave range.

Mirrors in space

Suppose that there was something, out in space, that would reflect radio waves. Waves from A that travel out into space could be reflected off it and return to earth, enabling stations such as B1 to receive A's signals. The situation just described is illustrated in **Figure 2**.

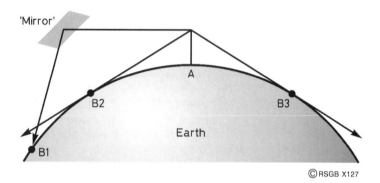

Figure 2 Reception of A's signals at B1 using the *sky wave*

©RSGB X127

Just such a mirror in the sky really *does* exist. It is not man-made, of course! The earth is surrounded by the atmosphere, a mixture of many gases, such as nitrogen and oxygen. The energy contained in the radiation from the sun is more than sufficient to *ionise* these gases, thus making them into electrical conductors. When a gas is ionised, some of its electrons are physically stripped out of the atoms and are free to move about, just as the electrons of a metal do in a wire. Consequently, we can regard this part of the atmosphere (the part illuminated by the sun) as acting like a sheet of metal, which *reflects* radio signals! It is not a perfect reflector, but is sufficient to produce long-range (DX) radio propagation via the *sky wave* under the right conditions. (A more down-to-earth example of ionised gases conducting electricity can be found in the fluorescent tube and the neon sign – many gases glow when they are continuously ionised.)

This conducting region at the extremity of the atmosphere is called the *ionosphere*, and it exists in layers between 60 km and 700 km above the earth's surface. When the ionosphere is sufficiently ionised, it glows; this is the natural phenomenon known as the *aurora borealis*, or the *northern lights*.

The property of the ionosphere that enables radio waves to be reflected does not act in a uniform way; it is very selective about which waves it reflects, and which waves go straight through it and into outer space. In general, it reflects only those waves with frequencies below about 30 MHz – the HF bands!

In Part 3 we will look at families of waves.

12 An LED flasher

Introduction

The LM3909 is an integrated circuit (IC) which will flash a light-emitting diode (LED). Using only two extra components and a battery, the circuit is cheap and has a very low current drain from a 1.5 V cell. The circuit can be used as a novelty flasher, an indicator for a dummy alarm bell box, or it could be attached to a torch so that it could be found easily in the dark! The simple circuit is shown in **Figure 1**.

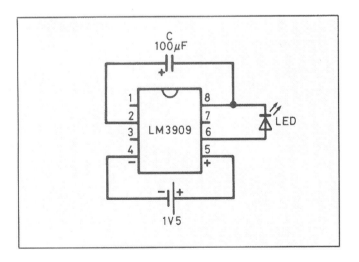

Figure 1 Circuit diagram of the LED Flasher. Pins 1, 3 and 7 of the IC are not used

Assembly

The circuit can be built on a small piece of Veroboard (the piece shown in **Figure 2** measures 15 holes by 10 strips). Using such a board, follow these instructions.

1. Depending on how far away you want the LED from the circuit board, solder a length of insulated wire to each lead of the LED. Use different colours of insulation – say, red and black, connecting the *red* lead to the *anode* (a) lead (the longer one) of the LED, and the black one to the *cathode* (k). Figure 2 shows these leads.
2. Cut the copper tracks as shown in Figure 2, using a 3 mm (⅛ inch) diameter drill, rotated between thumb and forefinger, or use the proper tool. Make *absolutely* sure that the tracks are completely broken!
3. Fit the IC holder in the correct position, using the cut tracks as guides, and make sure the small notch is facing towards the top of the board. Solder the pins to the copper tracks.
4. Mount the capacitor, positive end to the left, so that the positive lead is soldered to track F, which connects it to pin 2 of the IC; the negative lead is soldered to the right-hand side of track E, this being connected to pin 8 of the IC.
5. Solder on the battery leads, positive to the right, and the extended LED leads, positive downwards.

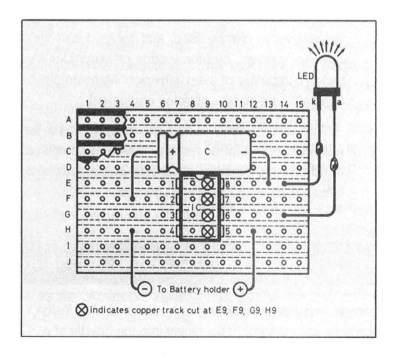

Figure 2 Board layout viewed from the component side. The tracks are cut under the board where shown

6. Check the circuit, and hold up the board to a bright light and look carefully for solder bridges between the tracks and pieces of copper swarf which may have escaped your inspection in 2 above! Remove whatever you find.

7. When all seems well, put the IC into the socket, ensuring that the notch or dot on the upper surface of the IC lines up with the notch on the holder. Line up each pin on the IC with the hole below it before pressing *gently* on the IC with the board supported on a firm surface.

8. Connect the battery; the LED should start to flash. The circuit is complete and working!

If you prefer, the whole circuit (battery included) can be mounted in a small plastic box, with the LED mounted on a clip and protruding through the panel. There are many other possibilities, and it is up to you to find an application for your own use.

Parts list

		Maplin code
LM3909	Integrated circuit	WQ39N
IC socket	8-pin DIL	BL17T
LED	5 mm diameter	WL27E
100 microfarad (μF)	Electrolytic capacitor (10 V)	FB48C
Battery holder	For AA-size cell	YR59P
Battery	1.5 V AA cell	

Small piece of Veroboard (15 holes by 10 strips)
Small plastic box (if required)
LED clip (if required)
Two lengths of coloured, insulated wire for LED (as required)

Availability

All parts can be obtained from Maplin Electronics Ltd.

13 Waves – Part 3

Introduction

When we talk about the spectrum being divided up into bands, this is just for our convenience; there are no natural divisions although, as we have seen, some of the properties of different bands really *are* different! Let's look at **Figure 1**, and see how the frequencies are divided up.

The divisions

- **Very low frequencies** (VLF) cover the range from a few kilohertz up to 30 kHz. Very long-range communication is possible, but at *very* small bandwidths. It is used for special purposes.
- **Long waves** (LW) are used for medium-distance commercial broadcasting and have frequencies from 30 kHz to 300 kHz.
- **Medium waves** (MW) are used for commercial broadcasting, and use frequencies from 300 kHz to about 1.5 MHz (1500 kHz). Typical range is about 200 km.
- **Short waves** (SW) encompass both the low-frequency (LF) and high-frequency (HF) amateur radio bands. There are nine narrow amateur bands in the SW spectrum between 1.8 MHz and 30 MHz. Some of these bands give round-the-world communication.
- **Very high frequencies** (VHF) span the range between 30 MHz and 300 MHz. Relatively short-range communication is possible. They were once used for broadcast TV before it moved to UHF. There are now three

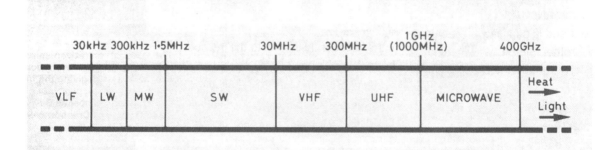

Figure 1 Diagram of radio frequency spectrum

amateur bands here – 6 m, 4 m, and 2 m. Repeaters are used to extend the usable range of mobile stations. VHF waves are not usually reflected by the ionosphere, but when they are, ranges of several thousand kilometres are possible. Weather affects these waves on a regular basis, however. In addition to amateur users, the VHF part of the spectrum is also used by the police, the fire and ambulance services, weather satellites, and many others,

- **Ultra-high frequencies**, sometimes called *centimetre waves*, cover the range from 300 MHz to 1000 MHz (or 1 GHz). The only amateur band in this range is the 70 cm band, and we share it with radar, TV and cellular telephone users as well.
- **Microwaves** begin at 1 GHz and extend to about 400 GHz. They are *never* reflected by the ionosphere, are partially attenuated by buildings, and are reflected from aircraft and cars. Microwave absorption in the atmosphere is quite significant, and rain and fog can attenuate microwaves quite heavily.
- **Heat, light** . . . Above 400 GHz we run into the infra-red bands and on into the visible light and ultra-violet bands. We generally take 400 GHz as being the limit of what we class as radio waves.

Bandwidth again

Complex signals need more bandwidth than simple signals. Even if it were possible, we would not be able to transmit a single television channel in the whole of the MW broadcast band! When TV used part of the VHF band, only five channels were possible in the range from 45 MHz to 68 MHz. By moving TV to the UHF band, we now have 47 channels between 470 MHz and 855 MHz!

It's your choice!

The number of permutations you have amongst all the modes and all the bands is enormous! Only you can decide what you are interested in and what you want to learn about. That is the attraction of amateur radio!

14 Choosing a switch

Simple, but be careful!

A switch is the simplest electronic component, but care is needed in choosing the correct one for the job. Here is a list of the main characteristics of a switch, which will help you to select what you want.

1. **Rating**. This gives the maximum voltage and current that a switch can handle. For example, a 250 V 1.5 A switch will switch mains voltage at a current not exceeding 1.5 A. If the current is greater than 1.5 A, the switch may get hot and fail. If the voltage is too great, the switch may arc each time it is switched off, thus wearing away the contacts.
2. **Number of poles**. A single switch can control many circuits; the number of poles tells you *how many* different circuits it can handle. See **Figure 1**.
3. **Number of throws**. This tells you the number of positions each pole can have. This is best illustrated in Figure 1. The simplest ON/OFF switch is a Single-Pole, Single-Throw (SPST) switch. A Single-Pole,

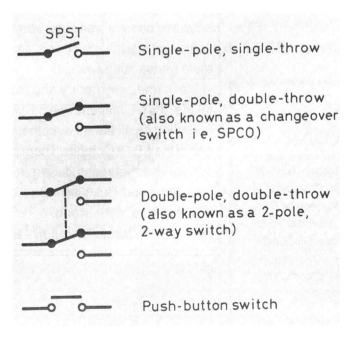

SPST

Single-pole, single-throw

Single-pole, double-throw (also known as a changeover switch i e, SPCO)

Double-pole, double-throw (also known as a 2-pole, 2-way switch)

Push-button switch

Figure 1 Some circuit symbols for different types of switch

Double-Throw (SPDT) switch may also be used as an ON/OFF switch, but is used mainly to change between two parts of a circuit, and is commonly known as a Single-Pole Change-Over (SPCO) switch. Two or more SPCO switches can be operated at once; Figure 1 shows an example. The two switches are said to be *ganged*. See also 'Types of switch' below.

4. **Number of ways.** When a switch has more than one throw, we tend to use the word 'ways' instead. This means that if a switch has one pole and six throws, we would normally call it a '1-pole 6-way' switch. Such switches tend to be rotary switches, as are described below.

Types of switch

(a) **Push-button.** These are found on calculators, telephones, electronic games and most equipment with a digital display.

(b) **Rotary.** These are switches controlled by a knob, and are turned instead of moved up and down. **Figure 2** shows the rear of such a switch and its circuit symbol. The commoner types of rotary switch are: 1-pole, 12-way; 2-pole, 6-way; 3-pole, 4-way; 4-pole, 3-way; 6-pole, 2-way. All these switches have 12 click positions, as you may have guessed, but each one comes with an adjustable end-stop so that you can set the correct number of ways according to the contacts on the switch.

(c) **Slide.** This switch is common on the cheaper types of radio, mainly as an ON/OFF or band-changing switch. They are not very rugged, but are small and cheap to produce. Very small types are manufactured for use on PCBs.

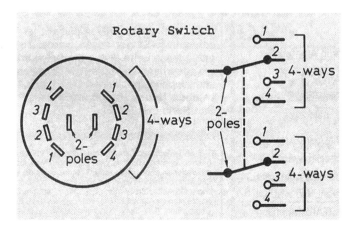

Figure 2 The rotary switch can select several different circuits at once

(d) **Toggle**. Available as two-way types or three-way, with a centre-off position. Can be manufactured to handle very high currents.

(e) **Micro-switch**. These are devices which are usually operated indirectly, such as when the cover is removed from a high-voltage power supply, or when the door of a fridge is opened. The 'micro' part of the name doesn't refer to the size of the switch, but to the *small* movement that is required to activate it.

15 An aerial tuning unit for a receiver

Introduction

Any length of wire will act as an antenna (or aerial) but, to get the best results from a transceiver or receiver, an aerial tuning unit (ATU) is required. This matches the *impedance* of your aerial to the impedance at the aerial socket of your radio. Impedance is like resistance, and is measured in ohms, but it is used for alternating currents, and hence is common in audio and RF engineering. Most receivers have an impedance at the aerial socket of about 50 ohms (Ω); aerial impedance, on the other hand, can be anywhere between 20 Ω to over 1000 Ω, depending on its length and its height above ground.

Fortunately, you don't *need* to be able to calculate your aerial's impedance; all you need is a device that will perform the matching operation for aerials with a large range of impedances, and this is exactly what is described here! The subject of *matching* is a complex one, but all you need to know is that most signals will become clearer, and that there will be less noise and interference. Stations will become louder, so you will probably be able to reduce the setting of your RF gain control (always a good thing to do).

This design of ATU covers all amateur and broadcast bands from 10 m (28 MHz) to 80 m (3.5 MHz), and is very easy to build. The circuit is shown in **Figure 1**.

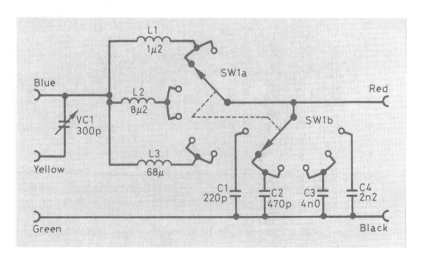

Figure 1 Circuit diagram of the Antenna Tuning Unit, showing the use of a 2-pole, 6-way rotary switch to select inductors (L) and capacitors (C)

Construction

1. Firstly, you will need a simple plastic case in which to house the ATU. The size should be approximately $85 \times 145 \times 50$ mm.
2. Start by drilling two 10.5 mm holes in the front of the case; these are for the 6-way switch and the tuning capacitor.
3. Drill three 8 mm diameter holes in the left-hand side of the box, for the three sockets, coloured blue, yellow and green.
4. On the right-hand side of the box, drill two 8 mm diameter holes for the red and black sockets.
5. Now, fit the 6-way switch (SW1), the tuning capacitor (VC1), and all the sockets to the case. Check that the vanes of the capacitor rotate smoothly when the shaft is turned.
6. Wire up the inductors (coils). Figure 1 and the wiring diagram of **Figure 2**, will help with this. As you can see, each one side of each coil is connected to two switch connections, the other end going to VC1.
7. Solder in the fixed capacitors. One end of each goes directly to the ground socket (black), and the other end goes to the switch.
8. Solder a wire between the green and black sockets. The output from the ATU comes from the red socket, and this is connected to the two tags in the centre of the switch, as Figure. 2 shows clearly.
9. Finally, connect the blue and yellow sockets to the tuning capacitor, and the ATU construction is complete.

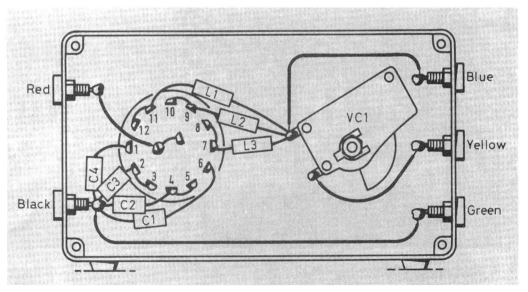

Figure 2 The internal view of the case shows the main tuning capacitor, VC1. This is a solid dielectric type, which has adjustable brass plates

In practice...

Figures 3 and 4 show two different ways of connecting your aerial to your ATU. In each case you will need to select each switch position in turn, and rotate the tuning capacitor through its full range while listening to a station. You should find that one switch position enables VC1 to produce a peak in the signal strength in the loudspeaker. At this point, your aerial and receiver are said to be *matched*. Stations in the same band will probably peak with VC1 at the same setting of SW1, but different bands will almost certainly require different positions of SW1.

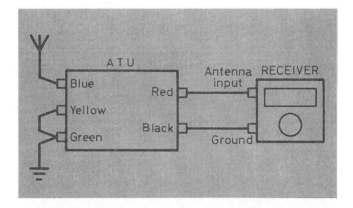

Figure 3 You may find that parallel tuning gives best results with your antenna

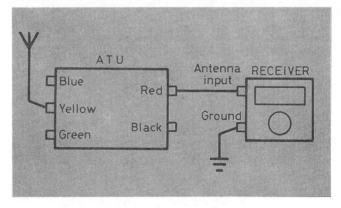

Figure 4 In other cases, series tuning could be the most effective arrangement

Parts list

Capacitors (all rated at 16 V or more)
- C1   220 picofarads (pF) polystyrene
- C2   470 picofarads (pF) polystyrene
- C3   1000 picofarads (pF) or 1 nanofarad (nF) polystyrene
- C4   2200 picofarads (pF) or 2.2 nanofarads (nF) polystyrene

Inductors
- L1   1.2 microhenries (μH)
- L2   8.2 microhenries (μH)
- L3   68 microhenries (μH)

Switch
- SW1   2-pole 6-way rotary

Sockets
- 4 mm type, one each of red, black, yellow, blue, green

Additional items
- Plastic or metal case, e.g. Maplin type YU54
- Two large knobs for SW1 and VC1

16 A simple 2 m receiver preamplifier

Introduction

Designed specifically to complement the modified air-band portable (also described in this series), this can be used with some success on many receivers suffering from 'deafness' on VHF.

The circuit

An RF preamplifier is a device which improves the input signal to an existing receiver, enabling it to work more effectively. Because of the noise which is added to the signal by the preamplifier, *very* weak signals may not be usefully enhanced, but stronger signals will be improved considerably.

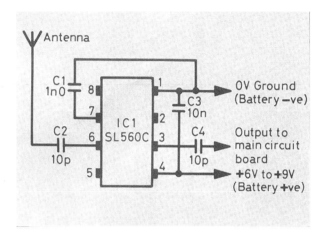

Figure 1 The integrated circuit is mounted upside down. Make sure you identify pin 1 which has a small dot next to it

This little circuit (shown in **Figure 1**) uses a GEC/Plessey integrated circuit type SL560C. With the addition of four capacitors, it is used between your aerial and the aerial input of the radio's PCB.

Putting it together

1. Use a small piece of prototype (matrix) board about 25 mm square. Use an 8-pin DIL socket for the integrated circuit (don't risk soldering the chip – it is seldom a risk worth taking). Figure 1 shows the connections to the socket, *looking from underneath*.
2. Make special note of the pin numbers, so that you know how to put the chip into the socket when you have finished. The positive and negative connections to the circuit are taken from the main PCB *after* the ON/OFF switch – so that the switch operates the preamplifier, too.
3. Unsolder the lead to the radio's telescopic aerial and connect it to the free end of C4, as shown. Then solder a short lead between the telescopic aerial and C2.
4. Sometimes it is possible to cajole your little preamplifier into the radio's plastic case, provided there is room and that you make sure that none of the soldered joints on your little board touch any of the metal inside the case when you replace the back and screw it on again.
5. If there is not enough room inside, then put the preamplifier into its own box, with battery and switch, and its own aerial. Keep the lead from the preamplifier to the aerial connection of the radio as short as possible – perhaps using screened cable.

Because your preamplifier is untuned, you will find not only that it helps with reception on 2 metres, but also that reception on the FM broadcast band is improved!

Parts list

Integrated circuit
    IC1   GEC/Plessey SL560C

Capacitors
    C1   1 nanofarad (nF) ceramic
    C2, C4   10 picofarads (pF) ceramic
    C3   10 nanofarads (nF) ceramic

Additional item
    Prototype broad approx. 25 × 25 mm

17 Receiving aerials for amateur radio

Introduction

For any radio receiver to work well, it must have some form of antenna, or aerial. In almost all domestic transistor radios, the aerial is built into the set, either as a ferrite rod (which looks like a rod of dark grey metal) on which are wound coils of wire, or as a chromium-plated telescopic metal rod. Some radios have both forms of aerial, using the ferrite rod aerial for long waves (LW) and medium waves (MW) and the metal rod for very high-frequency (VHF) stations using frequency modulation.

There's broadcast reception...

No aerial is perfect, and these two types are *far* from perfect! As in most mass-produced equipment, *they serve their purpose*, which is not critical, and they allow the radio to be carried around easily, because they are not big and bulky. The broadcast stations are very powerful, usually quite close, and the circuits in the radio are quite sensitive, so the need for large aerials disappears.

. . . and there's amateur DX

Here we have an application which, in comparison with the broadcast situation, couldn't be much more different. The stations do not use much power, they may be half the world away, and the requirement for good aerials and very sensitive receivers is paramount. The aerials must be large and they must be as high as possible, away from buildings and trees, which cause reduction in signal strength, and away from man-made sources of interference.

You may not have thought about this but, in general, the larger an aerial becomes, the longer the cable (or 'feeder') must be in order to reach your shack and the receiver inside it. Cables reduce the received signal, so what your aerial gains by being large, the feeder (if you're not careful) will lose!

A simple aerial . . .

We now know that a 'good' aerial is essential. But what is a 'good' aerial? It depends on your purse, your property and your enthusiasm! One of the simplest (and, incidentally, one which is *not* subject to the cable loss problem discussed above) is the **Long Wire**, shown in **Figure 1**. This is, quite literally, a long piece of wire going from a chimney stack to a tree or pole

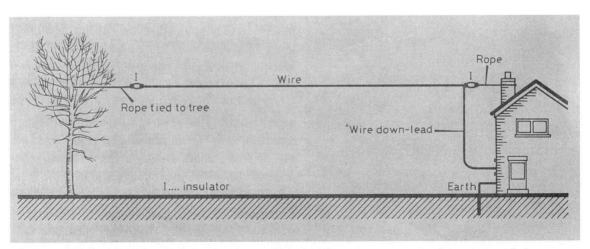

Figure 1 A tree at the bottom of the garden can provide a useful antenna support. Insulators can usually be obtained from Tandy stores or TV aerial suppliers

at the end of the garden. The longer it is, the better. Notice that the wire itself is *not* used to loop around the tree or chimney. Rope is used for both, and is secured to the end of the wire using an insulator of the 'dog bone' or 'egg' variety, to be found in profusion at rallies.

The wire is brought into the house through a window into the shack where the receiver is situated. The long wire aerial is best used with an aerial tuning unit (ATU), which is also described in this book. One of the advantages of the long wire is that it can be used on several frequency bands.

...and more complicated ones

Next up the ladder of complexity is the **dipole** (meaning two poles, or two elements). One form of dipole for lower frequencies would take the same basic form as the long wire, except that the feed to the receiver is taken, not from the end, but from the centre. The wire is essentially cut into two halves, and the two ends at the centre are connected to one end of a coaxial cable, which is then taken to the receiver. A smaller form of this is shown in **Figure 2**, which is conveniently mounted in a house loft. Aerials should always be mounted outside for best results, but will work when mounted inside, and the loft space is the logical situation.

The total length of the dipole should ideally be one-half of the wavelength of operation – hence the term 'half-wave dipole'. For example, a dipole for use on 20 m should be about 10 m long. Dipoles are thus 'single-band' aerials. They can be modified for use on several bands, and then become known as 'trapped dipoles', having coils and capacitors at certain points along their lengths.

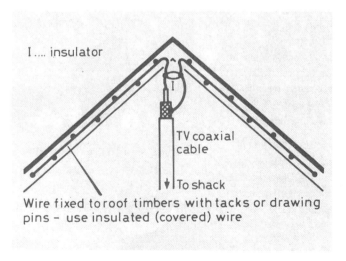

Figure 2 An indoor antenna should be mounted as high as possible

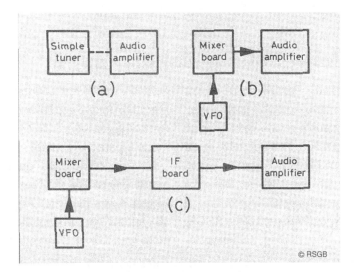

Figure 1 Build your Colt and watch it grow. A simple crystal set (a) becomes a direct conversion receiver (b) and finally an 80 metre amateur band superhet (c)

towards the input, and you can test what you have done stage by stage. You will see what this means as you progress with your construction.

Every receiver needs some audio frequency (AF) amplification to make the sound signals big enough for you to hear. The circuit for the AF amplifier is shown in **Figure 2**. It uses the TDA7052 integrated circuit (IC) plus a handful of extra components. R1 in conjunction with C2 and C3 *decouple* the battery supply, preventing any audio signals getting through to it and affecting other parts of the radio, when they are connected. C1 acts to prevent high frequencies (above the 3 kHz bandwidth) going into the amplifier input. A volume control, VR1, is connected across the amplifier input, so that the amplifier can accept signal inputs over a wide range. **Figure 3c** shows the connections, which are made with screened cable. The

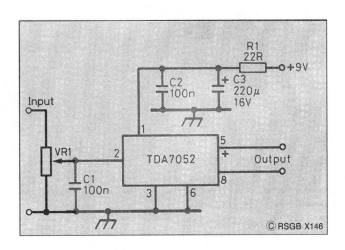

Figure 2 The Philips TDA7052 integrated circuit (IC) used in the audio amplifier needs very few extra components. It has a signal voltage gain of 100 times and the output is suitable for a loudspeaker or headphones

centre conductor of the left lead goes to the point marked 'input' in Figure 2, the braid being connected to the amplifier earth (0 V) tag. The right lead to VR1 goes to whatever signal source you have for testing – see later. VR1 is *not* mounted on the PCB.

The circuit is constructed on a small PCB or matrix board. Make sure that the electrolytic capacitor, C3, is soldered into the board the correct way – its positive and negative connections are shown in **Figure 3b** for reference. If you are **at all** concerned about soldering the IC into the board, enlist some help, *or*, obtain an 8-pin DIL socket, which you can solder in and then carefully insert the chip into the socket, making sure that is the correct way round. The markings on the chip are shown in **Figure 3a**.

The output leads from the amplifier go to a plastic 6.3 mm (¼ inch) mono jack socket, so that *neither* output lead is connected to the metal case. The amplifier will drive a pair of headphones or a small 8 Ω loudspeaker.

The battery leads must be the right way round also; the battery itself can be a PP3 or PP9, or you can use a small DC power supply.

Testing

First, check that all the components are in the right places, that your soldering is good, and that you have headphones or a loudspeaker connected. Set VR1 about halfway along its travel. Connect the battery.

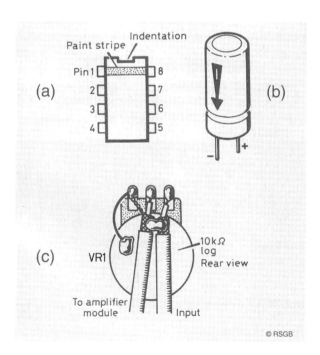

Figure 3 It is important to check the component connections carefully. The diagram shows (a) top view of the IC, (b) electrolytic capacitor and (c) volume control (VR1) connected across the input of the amplifier

A slight hissing noise should be heard; touching the input lead to the amplifier (the centre of the three connections on VR1) should produce a loud buzz. Touching the shaft at the same time will make the buzz quieter. This is the quickest way of confirming that your amplifier *seems* to be working. The only real test is to give it something meaningful to amplify! See the design of our *Crystal Radio Receiver* for full details.

Parts list

Resistors: all 0.25 watt, 5% tolerance
   R1   22 ohms (Ω)
   VR1   10 kilohms (kΩ) log

Capacitors
   C1, C2   0.1 microfarad (μF)
   C3   220 microfarads (μF) electrolytic 16 V

Integrated circuit
   IC1   TDA7052 audio amplifier

Additional item
   PCB   (see below)

Component suppliers:
   Maplin

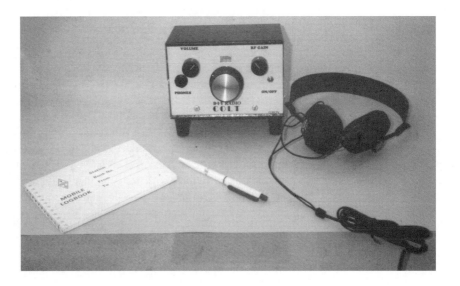

The next part . . .

The metal case will be marked out ready to receive the completed modules.

19 A crystal radio receiver

Introduction

Here's a quick project to fill in a winter's evening! It was originally designed as a piece of test gear for the *Colt* receiver, but can be successfully used by anyone as a first radio project. Another use for it would be as test gear for any audio amplifier project requiring an audio input for test purposes; this is ideal, as it does not require any power supply!

Details

The initial rough-and-ready test for the audio amplifier of the *Colt* receiver will have whetted your appetite; you will want to prove more conclusively that your amplifier works, and in a way that others in the household will appreciate. Buzzing noises are not convincing in this respect! You are going to put together a simple crystal set – the simplest type of radio that there is – and use it as a signal injector for your amplifier. In this way, you build a real medium-wave (MW) receiver which drives a loudspeaker, as an intermediate product of the construction of an 80 metre amateur band receiver!

The circuit diagram is shown in **Figure 1**. It has a 60-turn coil mounted on a small piece of paper or card wrapped round a ferrite rod. The coil has a connection made to its centre-tap (the middle turn of the coil). The tuning capacitor, VC1, is the most expensive part of the circuit but don't worry, it will be used in the final design of the receiver also! Connect the diode, D1, from the centre-tap to the input to the potentiometer of the amplifier circuit of the *Colt*. Be careful to connect the aerial to the *vanes* of VC1, and not to its *frame*, or you will experience some strange effects when you are tuning. If your amplifier is working correctly, you should be able to receive local stations on medium-waves quite well.

If you think the crystal set will be of use to you in the future as a signal injector, all you will need will be another variable capacitor! If you do not intend to use the amplifier, a small crystal earpiece will allow you to listen. Walkman-type headphones will *not* work!

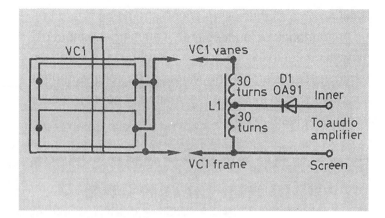

Figure 1 Both sets of moving vanes are joined as shown. L1 is wound on a ferrite rod with 32 SWG wire and centre-tapped. A single winding (no tap) can be used, joining D1 to the top of the winding

Parts list

Variable capacitor
VC1   125 + 125 picofarads (pF) twin-gang

Ferrite rod
About 100 mm to 140 mm long

Diode
D1   OA91 germanium diode

Additional items
Wire   2 m of enamelled copper wire, between 23 SWG and 32 SWG

Earpiece   High-impedance crystal type (only needed if you are not using the amplifier)

20 The varactor (or varicap) diode

Introduction

Many of the circuits for receivers and transmitters presented in this series rely upon the variable capacitor as a means of tuning. Another method of varying capacitance (without any moving parts) is provided by the *varactor diode*, sometimes called a varicap diode. This is a component which changes its capacitance as the voltage across it is varied.

The details

Figure 1 shows how a varactor diode might be connected to demonstrate its operation. Its symbol is that of an ordinary diode, with a capacitor symbol next to it. A variable voltage is applied across it in such a way that the diode is *reverse-biased*. This means that virtually no current passes through it – the positive voltage is applied to the cathode. Varactors are cheaper than variable capacitors, and they are tiny in comparison, very suitable for today's miniature circuits. If A and B were connected across the tuning coil in a simple receiver (with a series capacitor to block the DC from the battery reaching the coil), the tuning operation would be accomplished by turning the knob on the 10 kilohm potentiometer.

Varactors are available with different values, from less than 20 picofarad (pF) for VHF applications to 500 pF for medium-wave radios. They are

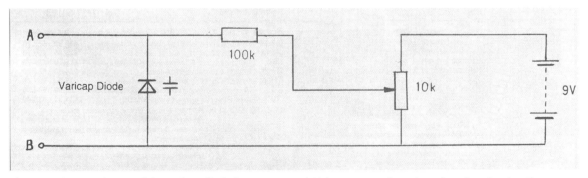

Figure 1 The capacitance of the varicap diode (between A and B) increases as the voltage is reduced, using the variable resistor

tuned usually by voltages between 2 V and 9 V. For a real application of varactors, you should consult the circuit diagram of the *Yearling* 20 metre receiver, elsewhere in this book.

In some circuit designs, several circuits are all tuned to the same frequency in order to improve the overall *selectivity* (the ability of the circuit to reject signals very close in frequency to the wanted signal). Special dual- and triple-varactors are available for circuits like this. Having been made at the same time from the same materials makes their individual characteristics virtually identical. Like all other diodes, they must be correctly wired into the circuit – their *polarity* is important.

Changes in temperature will cause the capacitance to change which, if it were part of an oscillator circuit, would cause the oscillator to drift – you would have to keep retuning the radio! This can be corrected by using a special integrated circuit called a *phase-locked loop* (PLL). Modern TV sets and satellite receivers use varactors and PLLs in this way.

Some useful varactor types

Type No.	Tuning range		Description
	pF/V	pF/V	
BB204B	42/2.0	15/12	Dual VHF
BB212	560/0.5	22/8	AM tuning
KV1235	450/2.0	30/8.5	Triple AM
KV1236	450/2.0	30/8.5	Dual AM
MV1404	120/2.0	9/10.0	HF tuning

21 A portable radio for medium waves

Introduction

The ZN415E integrated circuit (IC) can be used to make a very efficient AM portable MW broadcast radio with a built-in loudspeaker. Here's how!

The circuit

Figure 1 shows the circuit diagram of the portable radio. It's not as complicated as it may appear, especially after you have got started. L1 is a coil of wire mounted on a ferrite rod, acting as an aerial; VC1 is a variable capacitor which works, with L1, to tune in different stations. IC1 contains circuits of its own which boost the selected signal and it includes a *detector* which extracts the audio signal from the incoming RF signal. Earphones could be connected to the output of IC1 (between pins 4 and 5), but the output would not be powerful enough to drive a loudspeaker.

More sound

This is where IC2, an LM386 comes in. This is a small audio *power* amplifier which produces audio signals with enough power to drive a small

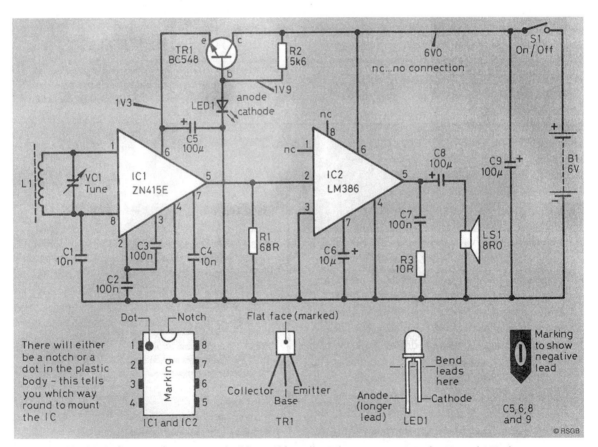

Figure 1 The circuit diagram of our easy-to-build portable radio. Take care to mount the ICs and LED the correct way round

loudspeaker, LS1. The radio uses a 6 volt battery, which is made by connecting four 1.5 volt AA cells in series ($4 \times 1.5\,V = 6\,V$) using a battery holder designed for this purpose. Although a 6 V supply is ideal for IC2, it is far too great for IC1, which needs only about 1.3 V. This lower voltage is provided from the 6 V supply by TR1 (an npn transistor), R2 and LED1 (a light-emitting diode). When current passes through an LED (see the description of the LED in this series) a reasonably constant voltage of 1.9 V appears between the anode and the cathode. Because of the voltage (0.6 V) that *always* exists between the base and emitter of a working transistor, the voltage on the emitter is about $1.9\,V - 0.6\,V = 1.3\,V$, and this is used as the power supply for IC1.

To keep the radio as simple as possible, no volume control has been fitted. Instead, you can use the directional properties of the ferrite rod aerial (see the information on ferrites in this book) to reduce the volume by rotating the set about a vertical axis using the handle provided.

Putting it all together

1. Start by covering the ferrite rod with Sellotape, or alternatively wrap a piece of paper tightly around it, and secure it with Sellotape. Then, with at least 2 metres of 24 SWG enamelled copper wire, wind 75 turns tightly around the rod. To be safe, leave about 50 mm of wire at the ends of the coil, then wrap the whole coil with Sellotape to hold the turns in place, leaving only the ends free. Then, using a small piece of sandpaper, remove the enamel from the last centimetre of each end of the coil.

2. Most of the components are mounted on a piece of Veroboard (the type with parallel copper strips on one side). The piece used on the prototype measured 32 holes by 10 strips, as **Figure 2** shows. Before you start fitting components, cut the copper strips as shown. It is easier to do it now than when the board is littered with components! The strips may be cut with a 3 mm ($\frac{1}{8}$ inch) twist drill rotated between thumb and forefinger. Resist the temptation to use a hand drill – the idea is just to cut the copper, **not** to drill right through the board!

3. Solder the IC sockets and the other components on the board as shown in Figure 2. Make sure that the IC sockets are fitted with their notches towards the *top* of the board, as viewed in Figure 2. Do not insert the chips yet. Always keep the wires left over from cropping resistors and capacitors, they will come in handy at times like this: make the wire links that are clearly shown in Figure 2. Connect the electrolytic capacitors (C5, C6, C8 and C9), the transistor and the LED the correct way round; then check it again when you have done it!

4. Finally, solder lengths of stranded insulated wire to act as 'flying leads' for future connection to L1, VC1, LS1, S1 and the battery connector.

5. Apart from the battery holder, everything is mounted on the case *lid*. This makes assembly and testing much easier, and eases fault-finding if

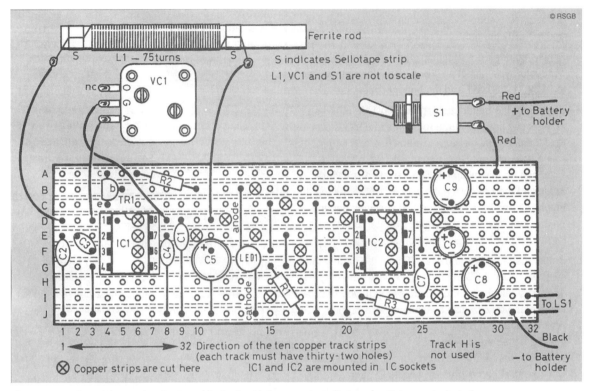

© RSGB

Figure 2 Veroboard layout for the Portable Radio. Make sure that all the wire links are included

the need arises! At the speaker position, make one large hole or a series of small holes to let the sound out. The ferrite rod may be stuck to the lid, as may the loudspeaker. Drill holes of the correct size to fit the particular types of variable capacitor (VC1) and switch (S1) that you are using. The Veroboard may be held in position by Blu-Tack or double-sided sticky tape.

6. Before inserting IC1 and IC2, connect the battery and switch S1 on. The LED should glow dimly (you may have to shield it with your hand in order to see it). If you have a test meter, check that there is about 1.3 V between pins 6 and 4 of IC1. If the reading is around 6 V or there is no glow, you may have connected the LED the wrong way round! When everything seems normal, switch off and disconnect the battery. Insert IC1 and IC2, making sure that the pins are straight and lie immediately above their corresponding holes in the sockets, and that the notches line up with the notches in the holders. Then push *gently* downwards on each IC in turn until the chip is firmly seated in its socket.

7. Switch on! By rotating the tuning capacitor, VC1, you should now be able to tune in many stations, rotating the radio to give you some volume control.

Final touches...

The handle was made with part of an old leather belt, secured to the case with 'number plate' nuts and bolts from Halfords. The loudspeaker grille is the lid from a *pot-pourri* container, and some extra holes were drilled in the back to improve the sound. See what you can find to finish off your radio!

Parts list

Maplin order codes are given for most of the parts, but you should get used to using the 'beg, borrow or steal' technique, or to use your ever-expanding junk box.

		Maplin code
Resistors: all 0.25 watt, 5% tolerance		
R1	68 ohms (Ω)	M68R
R2	5.6 kilohms (kΩ)	M5K6
R3	10 ohms (Ω)	M10R
Capacitors		
C1, C4	10 nanofarads (nF) or 0.01 microfarad (μF) ceramic	BX00A
C2, C3, C7	100 nanofarads (nF) or 0.1 microfarad (μF) ceramic	YR75S
C5, C8, C9	100 microfarads (μF) electrolytic, at least 10 V	FF10L
C6	10 microfarads (μF) electrolytic, at least 25 V	FF04E
Semiconductors		
IC1	ZN415E radio chip	
IC2	LM386 audio power amplifier	UJ37S
LED1	3 mm green LED	WL33L
TR1	BC548 npn transistor	QB73Q
Additional items		
LS1	Miniature 8 ohm loudspeaker	WB08J
S1	Miniature SPST toggle switch	FH97F
Ferrite rod	Length approx. 100 mm	YG20W
	24 SWG enamelled copper wire	BL28F

Additional items (*continued*)

VC1	Tuning capacitor 140 to 300 picofarads (pF)	FT78K
	Tuning knob	FK41U
	8-pin DIL IC sockets (two required)	BL17T
	4 × AA-size battery holder (long)	HF94C
	PP3-type clip for battery holder	HF28F
	Plastic box approx. 158 × 95 × 54 mm	LH51F
	0.1 inch Veroboard, min. size 32 holes × 10 strips	JP46A

Plus
Stranded insulated conductor for flying leads
Multicore solder
Materials for handle and speaker grille
Double-sided sticky tape or Blu-Tack
Sellotape
Glue
Four AA-size 1.5 V batteries

22 The Colt 80 m receiver – Part 2

Introduction

In Part 1 we constructed the audio amplifier module for the system and tested it in a very simple way. If you did what was suggested and built the simple Crystal Set to use as a signal source, you will know just how well the amplifier works.

The case

Metal cases for the project are available from Maplin, telephone 01702 554 161 (code XB67). From the photograph on p. 72 you can see the way the components are mounted. The audio amplifier is seen at the top right of the base, to the right of the tuning capacitor VC1. The next in this series will deal with the variable-frequency oscillator (VFO) and VC1. The current part deals with preparing the case to receive the components.

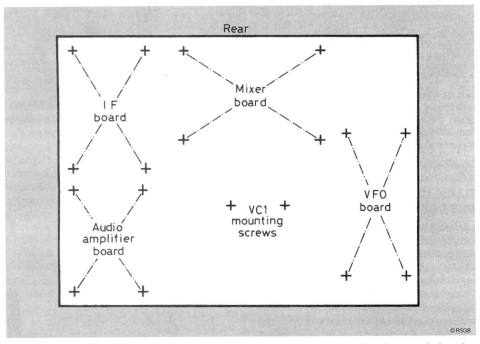

Figure 1 Fixing holes for each module are best measured from each printed circuit board or matrix board

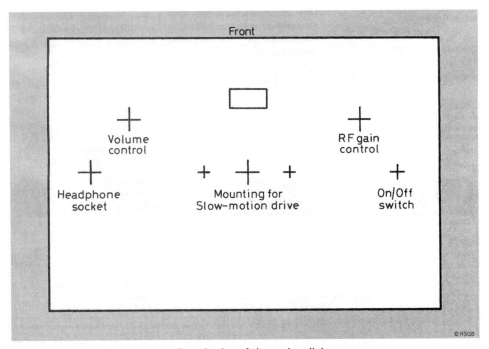

Figure 2 Position the slow motion drive to allow viewing of the tuning dial

Figures **1** and **2** show the markings for preparing the front panel and base. All the circuit boards and the tuning capacitor are mounted on the base using 10 mm stand-off pillars with 6BA bolts. The board locations are shown in Figure 1. The front panel control positions are shown in Figure 2, together with the small rectangular hole for viewing the tuning dial.

The best way to mark out the holes for the boards is to lie the boards flat on the base (before you've started soldering the components in) and marking the base through the holes in the boards. This minimises the scope for errors!

A *reduction drive* is used between the tuning knob and the capacitor shaft. This is simply a gear mechanism that slows down the capacitor shaft by a factor of six compared with the tuning knob, and makes tuning very much easier. The recommended variable capacitor also has a pulley wheel mounted on the shaft. Glued to this wheel will be a scale marked with frequency and is visible through the rectangular hole in the front panel.

The next part...

The variable-frequency oscillator and mixer will be added to the project.

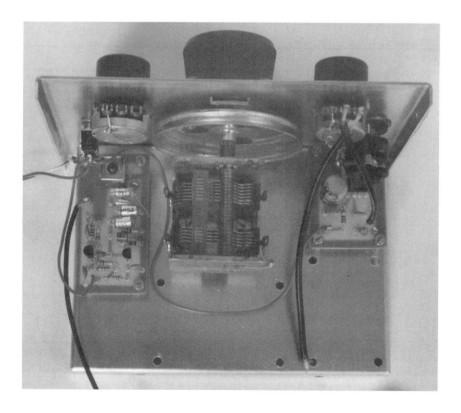

23 A simple transistor tester

Introduction

Although transistors aren't used as much as they were before integrated circuits came along, a transistor tester is still a useful piece of test equipment to have around the shack. This design is about the simplest possible and will produce an indication of whether a transistor is giving any current gain; this does not necessarily mean that the transistor is perfect but that it is working to a certain extent. This tester will **not** test field-effect transistors (FETs). If you buy a bag of transistors at a rally, this tester is useful for giving a yes/no indication of which ones go straight in the bin and which are kept for further use.

How it works

Figure 1 shows the simple wiring circuit. In order to explain the working of the circuit, a circuit diagram is shown in **Figure 2**, with the npn transistor, TR1, under test shown as part of the circuit. The pnp/npn selector switch, SW1, is omitted for clarity.

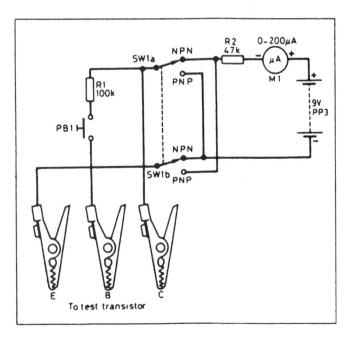

Figure 1 Circuit diagram of the transister tester

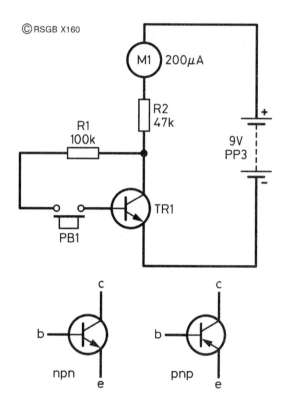

©RSGB X160

Figure 2 Testing an npn transistor

Any current that flows through the transistor, TR1, must flow from the battery, through the meter, M1, and through the protective resistor, R2. R2 prevents excessive current flowing through the meter and damaging it. Even if there is a short-circuit between emitter and collector, the maximum current that will flow is given by the simple equation

$$I = \frac{V}{R},$$

where I is the current flowing in amps,
          V is the battery voltage, and
          R is the total circuit resistance in ohms.

Putting in the correct values, gives

$$I = \frac{9}{47\,000} = 0.000191 \text{ A, or 191 microamps } (\mu\text{A}).$$

The resistance of the meter, M1, will cut this down a little more, but it is within the indicating range (200 μA) of the meter.

As shown, with the push-button switch, PB1, *open*, a good transistor will not draw any current from the battery, and M1 will thus remain at zero. When PB1 is *pressed*, a *very small* current is injected into the base of the transistor. If the transistor is working, it will produce a much larger current between the collector and emitter, and this current will also flow through M1 and R2, giving a significant reading on M1, showing that all appears to be well. If an appreciable current flows when PB1 is *open*, then your transistor is suspect.

Don't be put off by the apparently complicated switch, SW1. It is there to allow the other type of transistor, the pnp type, to be tested. All it does is reverse the battery connections, so that the emitter goes to the *negative* battery terminal for testing an npn transistor, and to the positive terminal for a pnp type!

Most transistors in common use are of the npn type, which is why Figure 2 shows the testing of an npn type. The connections to the different transistor *encapsulations* (shapes) are given in any good component catalogue. Avoid the trial-and-error method to discover the connections to a transistor. This is unscientific, and can be very frustrating, particularly if the transistor is faulty in the first place!

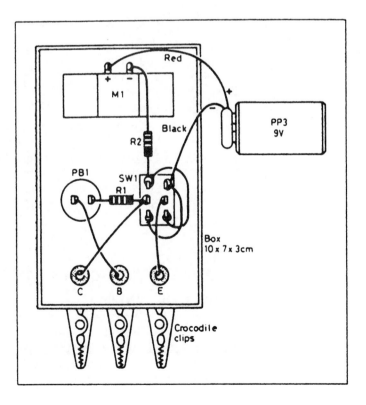

Figure 3 This shows how to wire up the tester

Construction

The unit can be built into a small plastic box, the components being soldered directly to the fixed terminals of the meter and the two switches; no circuit board is necessary! Use different colours of wire for the three test leads, and make sure you know which is which! Check the connections against the wiring diagram of **Figure 3**. When you are confident that all is correct, connect the battery and make sure there is no reading on the meter when nothing is connected to the crocodile clips!

Find any transistor for which you know the connections and the type (npn or pnp). Set the npn/pnp switch accordingly. Connect the three clips, making sure that they do not touch each other. A small reading on the meter at this stage means the transistor is suspect; a large reading means it is not working and should be thrown away! Press PB1 and watch the meter; a reading greater than half of full-scale indicates a good transistor. If it is less than half, you may have a transistor with 'low gain'; it may be usable for non-critical applications, but if you are in any doubt – throw it away!

Parts list

Resistors: all 0.25 watt, 5% tolerance
    R1   100 kilohms (kΩ)
    R2   47 kilohms (kΩ)

Additional items
    PB1   Push-button switch
    SW1   DPDT (double-pole double-throw) switch
    M1   Micro-ammeter – *not* less than 200 µA full-scale deflection
            Crocodile clips (3 needed)
            PP3 battery and connector
            Plastic box about $10 \times 7 \times 3$ cm

Common types of transistor

BC108, BC109, 2N2369A

These are small-signal npn types, used in audio amplifiers. They are in metal cases (called TO18) and have a tab next to the emitter lead. Common types have a B or C suffix (e.g. BC109C). The C suffix indicates a higher current gain than those with a B suffix. The 2N2369A is specially designed for radio use at high frequencies.

2N3703, BC212L, BCY71

These are pnp transistors, and so must be used with the collector and base *negative* with respect to the emitter. These three types are used in small amplifiers and audio oscillators. The first two have plastic encapsulations (TO92), while the BCY71 has a metal case (TO18).

BFY50, BFY51, BFY52

For slightly higher powers, these are ideal. They have been used in novice transmitters up to 600 milliwatts (mW). The TO5 case is a scaled-up version of the TO18 case. They are all npn types.

2N3055, 2N3773, TIP35C

These are high-power transistors in bigger encapsulations. The thick metal TO3 case of the 2N3055 and 2N3773 is designed to bolt to a heat sink, a large piece of metal which conducts the heat into the air more rapidly than the transistor itself can. The TIP35C is made of plastic but has a thick metal tab by which it, too, can be bolted to a heat sink.

24 An introduction to transmitters

Introduction

We usually think of a transmitter as being a 'black box'. However, that is the form a transmitter takes for our use on the amateur bands. Many electrical circuits are transmitters, even though transmitting may not be their primary function!

Anything that emits electromagnetic energy at *any* frequency is a transmitter, from radio at the low-frequency end of the spectrum, to gamma rays at the high-frequency end. We all know that a magnet will attract certain metals and that a comb rubbed on your coat sleeve will pick up small pieces of paper. The former is an example of the effect of a magnetic field, the latter of an electric field. Electromagnetic fields are combinations of both types of field, and are produced whenever an electromagnetic wave is transmitted.

What frequency?

Many everyday objects have a natural frequency of oscillation. This is called their *resonant frequency*. A wine glass will ring when struck gently; an empty wine bottle will sound if you blow across the top; a guitar string will vibrate when plucked. These are all examples of resonance, and the resonant frequencies will **not** change unless the objects themselves are changed physically in some way. These are resonances in *sound*; we are primarily interested in electrical resonances.

The basic electrical resonant circuit is the combination of an inductor (coil) and a capacitor, as shown in **Figure 1**.

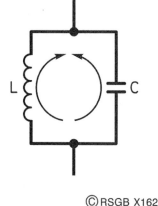

Figure 1 The basic electrical resonator. The energy in the circuit alternates between the inductor and the capacitor

©RSGB X162

A pulse of energy applied to this *tuned circuit* will make it *ring* (oscillate) at its resonant frequency. The energy in the circuit transfers between the inductor and the capacitor every cycle of the oscillation. Just like the wine glass, its oscillation dies away because it is losing some energy to its surroundings – it is transmitting! The frequency of the resonance depends on the values of L and C.

Keeping it going

If we want to keep the circuit oscillating, rather than having it die away, we must supply the circuit with just enough energy to replace the energy lost both by radiation and by losses in the circuit itself. Because of this, you will find in all oscillator circuits, a transistor, valve or FET working with the tuned circuit to provide this extra energy.

As it stands, of course, even with its transistor, our oscillator will not radiate very far. Connecting an aerial to it, and a Morse key to interrupt the power supply, it would become a very low-power CW transmitter. Add a couple more transistors to form a radio-frequency (RF) amplifier, and you have the basis of a simple low-power (QRP) transmitter.

Resonant circuits can also be made using quartz crystals; these work at the crystal frequency only, and this is marked on the crystal case.

A tiny spark transmitter

This is a simple piece of test gear that will increase your knowledge and understanding of resonance. You can use it to estimate the resonant frequency of most of the inductor/capacitor (LC) tuned circuits that you build. The circuit is shown in **Figure 2**. It operates around a relay. Any relay that operates from a 6 V to 9 V source and has contacts which are normally closed (i.e. closed when the battery is not connected). Fit the relay, a toggle switch and the battery in a metal box, and connected up as shown in the diagram. Some foam rubber inside the box may help to reduce the escaping noise of the relay. A small hole in the side of the box enables the 2-turn loop to emerge. This should be about 40 mm diameter, made with insulated wire. Switch on; there should be a loud buzzing noise from the relay. If not, you have probably chosen the wrong contacts on the relay!

When it is working, bring the loop close to the aerial of a radio – it should produce a loud noise from the speaker!

How it works

When you switch on, current flows through the relay contacts and through the relay coil. The relay operates and opens the contacts, causing the relay to 'drop out'. When it does, the circuit is completed again and the contacts are opened, and the cycle repeats. Each time the relay contacts open, there is a small spark between them, causing very rapid current surges through the wire loop. This makes the loop transmit RF energy, very briefly. In the early days of radio, this type of circuit was known as a *spark transmitter*.

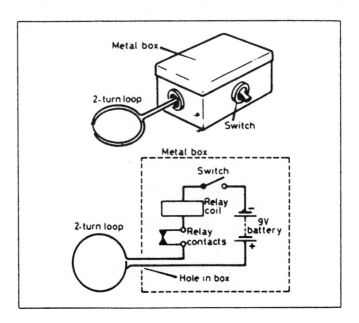

Figure 2 The current path is interrupted when relay is energised as shown above

Make a tuned circuit

Use a discarded toilet-roll centre, and wind about 10 turns of enamelled copper wire round it, keeping each turn close to the next. Scrape off the enamel for about 1 cm at each end, and solder a 100 picofarad (pF) capacitor (or a variable capacitor of about the same value) between the ends. The resonant frequency should be about 10 MHz. If you have used fewer turns or a smaller capacitor, the frequency will be higher.

Measuring the resonant frequency

Set up the buzzer as shown in **Figure 3**, with the loop around one end of your coil. Then make a similar loop, solder it to the end of a piece of coaxial cable going to the aerial socket of a calibrated receiver. Set the buzzer going, tune the receiver around 10 MHz, and search for the maximum noise level from the speaker. When you have found it, move the two loops as far away as possible from the main coil. This is called reducing the *coupling* between the coils, and it may result in a slightly different, but more accurate, resonant frequency.

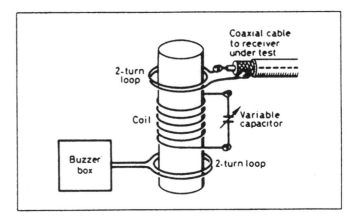

Figure 3 Experiment with the spacing between the loop and the tuned circuit

Parts list

Any small relay which operates between 6 V and 9 V
Metal box – do not use a plastic box!
9 volt battery and connector
On/off (SPST) toggle switch
Plastic foam, as required

25 The Colt 80 m receiver – Part 3

Introduction

In Part 2 we marked out the case ready for installation of the modules when they are completed. In this part, the building of the variable-frequency oscillator (VFO) and mixer will be described. This will produce a type of receiver known as *direct-conversion*, because it converts the radio-frequency (RF) signal directly into an audio-frequency (AF) signal which we can hear in a loudspeaker after amplification. A block diagram of the system is shown in **Figure 1**. The modifications needed to make a full superheterodyne receiver will be left until later.

The direct-conversion receiver covers the 80 metre amateur band and will receive both Morse (CW) and speech (SSB) signals. The audio amplifier was covered in Part 1, so your *Colt* is rapidly taking shape! By the time your construction has reached the end of this part, you will have a receiver ready to use, even if the project is not yet complete!

The direct conversion process

Like most things in radio, the principles of direct conversion are not difficult. From the aerial, the signal we want to hear is selected by the *tuned filter*, which rejects the signals we don't want. The signal then enters the

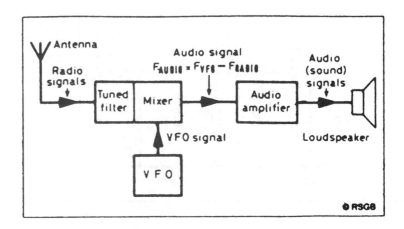

Figure 1 Stages of direct-conversion receiver

mixer, along with the signal from the *VFO*. The VFO produces a sine wave whose frequency can be varied across the whole of the 80 metre amateur band (3.5–3.8 MHz), by turning the knob on the tuning capacitor, VC1. The mixer produces, at its output, *two* signals; one signal is at the frequency of the **sum** of the signal and VFO frequencies, the other is at the **difference** of the two frequencies. It is the latter that we want. Let's look at the numbers involved. If the signal is at 3650 kHz and the VFO is at 3651 kHz, then the sum frequency is 7301 kHz, and the difference frequency is 1 kHz. If we feed the output of the mixer into our audio amplifier, the 7301 kHz signal is automatically removed (it is far too high to be considered an audio signal!) and the resulting 1 kHz signal is amplified and fed to the speaker, producing a note which we can hear!

Building the VFO

Figure 2 shows the circuit of the VFO. It is a tried and tested circuit, and should work first time. It uses a *field-effect transistor* (FET) for TR1, the oscillator itself. RFC is a *radio-frequency choke*, a coil of wire which will pass a direct current (DC) but which will prevent radio-frequency (RF) signals getting through.

All VFOs have a tuned circuit which, in this case, is formed by the coil L1 and the capacitors C1 and VC1. The frequency will also be affected to some extent by C2, C3, C4 and C5. Transistor TR2 is an *emitter follower*, a stage which gives no voltage gain but provides a good *buffer* stage, isolating the VFO from the effects of the stages that follow it. When building a VFO, the

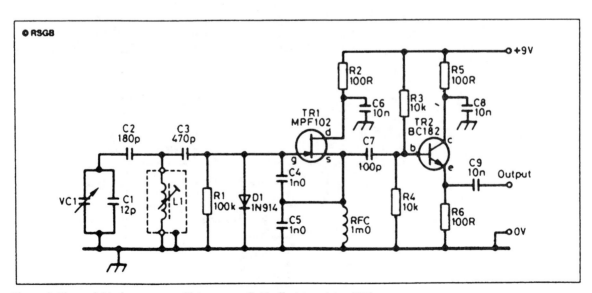

Figure 2 The variable-frequency oscillator uses a field-effect transistor (FET)

parts must be securely mounted. If components move, so does the frequency! At worst, the oscillation will become unstable and the VFO will be useless. Keep the component leads as short as possible – this improves their mechanical stability as well as their electrical stability!

Mount the parts on the printed-circuit board (PCB) or matrix board and, when completed, the VFO should look like the one shown on the left in the photograph on p. 72. Make sure that TR1, TR2 and D1 are the right way round.

On completion, check the component positions then mount it in the case as shown in the photograph, to the left of the tuning capacitor when viewed from the rear. The VFO coil, L1, will need some adjustment, but that will have to wait until the mixer is built. Connections to the other boards are made with screened cable.

The mixer board

So far, we have an audio-frequency (AF) amplifier and a VFO; the addition of a mixer board gives us a complete direct-conversion receiver for 80 m. The mixer circuit diagram is shown in **Figure 3**. Let's follow the signal path.

- From the aerial, the RF signal goes to the gain control potentiometer, RV1. This reduces very strong signals, to prevent them overloading the mixer.
- To select the required band of signals, a *bandpass* filter is made up of RF transformers T1 and T2; these are tuned by C1 and C3, and are coupled together by C2.

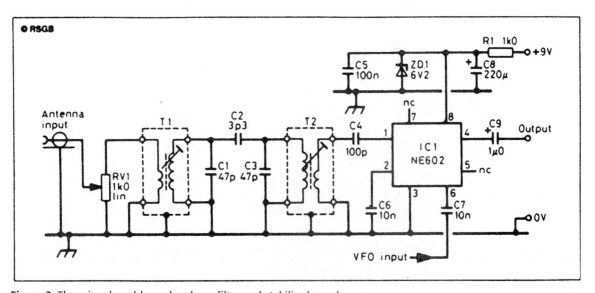

Figure 3 The mixer board has a bandpass filter and stabilised supply

- After the filter, the signal is coupled into the integrated circuit (IC) mixer type NE602, by the capacitor C4. Capacitors C5 and C8 *decouple* the supply line, to prevent unwanted signals on the supply from disturbing the VFO operation. The use of the term *decouple* is exactly the opposite of *couple*; when two circuits are *coupled* together, the signal passes from one to the other; when two circuits are *decoupled*, signals **cannot** pass from one to the other.
- The NE602 works with a 6 V supply; it is produced here from the 9 V supply by the *Zener diode* ZD1 and the resistor R1. ZD1 operates at 6.2 V, and gives a steady output for the mixer. The audio output from the mixer appears at pin 4 or IC1. This is taken to the audio amplifier board via C9 and the volume control (see Part 1).

Care must still be taken to insert some components the right way round. These are the electrolytic capacitors, C8 and C9, the Zener diode, ZD1, and the integrated circuit, IC1. Check all component positions and make sure all your soldered joints are bright and shiny.

Putting it together

The interboard wiring, shown in **Figure 4**, uses screened cable; ideally, this should be thin coaxial cable, but screened microphone cable is suitable. The diagram shows how the two controls, the RF Gain and Volume, are connected to the boards. The leads marked '+9 V' are all connected to the battery supply via a miniature on/off toggle switch. Double check all connections before connecting the battery.

Setting the VFO

Very little adjustment is needed to get the receiver going. Firstly, the VFO must be adjusted to cover the required band, in this case 3.5–3.8 MHz.

If you have a frequency counter, connect it to the output of the VFO. If you haven't, read this part anyway so you understand the process, then another means of setting the VFO will be given especially for you! Rotate VC1 until the vanes are fully meshed. Very carefully, adjust the core of the VFO coil (L1) with a plastic trimming tool so that the frequency approaches and settles at 3.500 MHz. When you rotate VC1, the frequency should increase to **at least** 3.800 MHz at the far end of its travel.

In the absence of a frequency counter, borrow a communications receiver, set it for SSB reception (USB or LSB) on exactly 3.500 MHz. Set VC1 with the vanes fully meshed and turn the core of L1 in both directions until you hear a whistle in the communications receiver. Rotate the core so that the whistle reduces in frequency. It will eventually fade out at around 200 Hz;

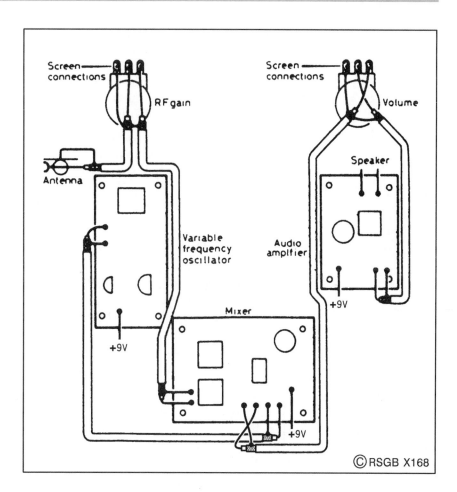

Figure 4 Make sure that interconnections between the boards are correct, including cable screens

turn the core a little further and then leave it at that position. Rotate VC1 right to the other end of its travel, and search for the whistle with the tuning knob of the communications receiver. Check that the VFO frequency is at least 3.800 MHz.

Then, whichever method you are using for frequency measurement, mark the dial with frequency steps of 50 kHz. Setting and calibration are finished!

Setting the mixer

Again, there are various ways of doing this. If you have a signal generator, inject a signal at a frequency within the 80 m band, and adjust the cores of T1 and T2 sequentially for maximum output.

If you haven't a signal generator, connect an aerial to the mixer input, set the RF Gain to maximum (fully clockwise), and find a consistent signal. Adjust the volume control to a comfortable level. Rotating the core of T2 with your trimming tool, maximise the output. Then do the same with T1, although this will have much less effect. Find another station, and check that the positions of the cores aren't too different for a maximum signal.

You may find that your receiver benefits from the insertion of an aerial tuning unit (ATU) between the aerial and the input, to compensate for the impedance of your aerial not being $50\,\Omega$. A design for such an ATU is presented in another part of this series. If the signals are still weak, connect the ATU to the junction of C1 and C2 via a 100 pF capacitor.

Try listening!

Remember that 80 m is a variable band. During daylight hours, your will hear Morse signals at the lower end of the band, and some British and closer continental stations between 3.7 and 3.8 MHz. In the evenings, stations up to 1000 miles away should be heard. Look for Novices around 3.7 MHz!

Parts list – VFO board

Resistors: all 0.25 watt, 5% tolerance

R1	100 kilohms (kΩ)
R2, R5, R6	100 ohms (Ω)
R3, R4	10 kilohms (kΩ)
RV1	1 kilohm (kΩ) linear

Capacitors

C1	12 picofarads (pF) polystyrene
C2	100 picofarads (pF) polystyrene
C3	470 picofarads (pF) polystyrene
C4, C5	1 nanofarad (nF) polystyrene
C6, C8, C9	10 nanofarads (nF) polystyrene
C7	100 picofarads (pF) min. ceramic
VC1	140 + 140 picofarads (pF) variable

Semiconductors

TR1	MPF102	FET
TR2	BC182	npn
D1	1N914	silicon

Inductors

L1	Toko KANK3334	
RFC	1 mH	RF choke

Parts list – mixer board

Resistors: 0.25 watt, 5% tolerance
    R1   1 kilohm (kΩ)
    VR1   1 kilohm (kΩ) potentiometer (linear)

Capacitors
    C1, C3   47 picofarads (pF) min. ceramic
    C2   3.3 picofarads (pF) min. ceramic
    C4   100 picofarads (pF) min. ceramic
    C5   100 nanofarads (nF) min. ceramic
    C6, C7   10 nanofarads (nF) min. ceramic
    C8   220 microfarads (μF) electrolytic 16 V
    C9   1 microfarad (μF) electrolytic 16 V

Integrated circuit
    IC1   Philips NE602 or NE602A

Additional items
    T1/T2   Toko KANK3333
                On/off switch
                Miniature toggle switch

The next part...

The IF amplifier and the Beat-Frequency Oscillator will be added to convert the *Colt* into a superheterodyne receiver.

26 A two-way Morse practice system

Introduction

It is said that the pleasure of sending and receiving Morse code more than compensates for the learning effort needed. Transmitters for Morse code can be much simpler than those needed for any of the speech modes.

Many people choose to learn sending and receiving on their own. It can be much more fun if you have someone to learn with you, and it is for this reason that the following project arose. It comprises just one small circuit, and you build an identical circuit for each of the people who want to learn with you. All the individual circuits are connected with their output leads in *parallel*, as the two-way circuit of **Figure 1** shows.

A simple circuit

This diagram shows two identical circuits, as would be used if two people wanted to learn together. The circuit centres around our old friend the NE555 integrated circuit (IC). As it is connected here, it works as an *astable multivibrator*, a daunting name for what is essentially an oscillator. Each circuit is self-contained, having its own battery, Morse key, sounder and

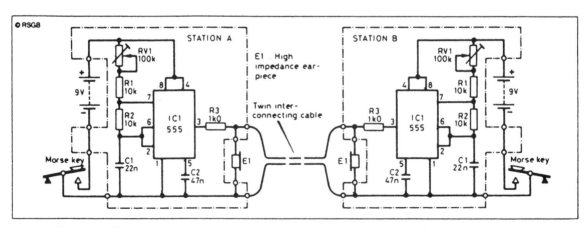

Figure 1 The circuit diagram shows two stations, but you could connect several more if needed

plastic box. The Morse key makes and breaks the power supply to the circuit, thus turning on and off the note emitted by the earpiece. The frequency of the note is controlled by the variable resistor RV1, resistors R1 and R2, and the capacitor C1. The range given by RV1 is approximately from 500 Hz to 2200 Hz.

R3 protects the output of the IC against accidental shorting either to the 0 V or to the 9 V supply rails. For best results, a crystal earpiece or high-impedance headphones should be used. You could try Walkman-type headphones, but the volume may be too low for you.

The units may be interconnected with thin twin cable (the sort used for wiring doorbells) and can be comfortably separated by 25 m, more than enough to communicate between rooms, or even with a neighbour!

In use

When an operator presses his key, the note is heard by himself and by everyone else who is connected. If more than one key is pressed at the same time, two or more notes are heard by everyone! It helps if each person adjusts his own potentiometer, RV1, to give a note which is different from the others. When one operator stops sending, the other can start immediately, without pausing to change from transmit to receive. In practice, this technique is known as *full break-in*.

The current drawn by the circuit is only 10 milli-amps (mA) when the key is down, meaning that the life of a typical PP3 battery will be virtually its shelf-life! No switch is needed, because the key acts as the switch. The current drain will be even less if the NE555 is replaced by an ICM7555 IC. It has exactly the same pin connections as the NE555, so no circuit modifications are needed.

The circuit board

The circuit is constructed easily on a small piece of Veroboard of the copper-strip type. **Figure 2** shows the layout of the prototype, the board measuring 18 holes by 10 strips. **Be aware that there is no row 'I' when you are transferring mental images of where the parts are to real positions on the board!** Break the copper tracks where shown with a 3 mm ($\frac{1}{8}$ inch) twist drill, rotated carefully between thumb and forefinger. Hold the board up to a bright light to make sure that the tracks are completely broken and that there are no fragments of copper swarf shorting adjacent tracks. Then, solder in the wire links shown in Figure 2, followed by the resistors, RV1, and the capacitors, C1 and C2. Solder the battery connector leads to tracks A and G, G being positive. Use insulated wire to connect from the board to the jack sockets used as connectors for the key and the earphones. If you use

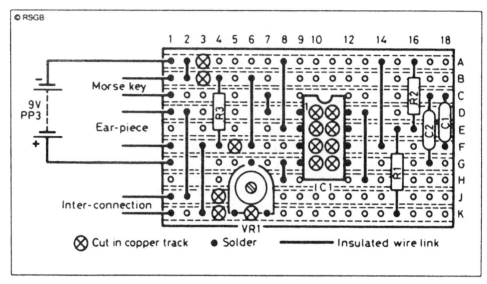

Figure 2 The Morse Duet is an easy project to build on stripboard (Veroboard)

different sized jack sockets (i.e. 2.5 mm and 3.5 mm) you won't plug the earphones into the key socket! A two-screw terminal strip is useful to connect the cable runs between operators.

The key

The circuit is designed to be used with a *straight key* (one that moves up and down). It is a requirement of the UK Morse test that a straight key is used, so it is very sensible to learn sending with a straight key before you try anything more complex! You will find many to choose from at rallies.

Parts list

		Maplin order codes
For each board, you will need:		
Resistors: all 0.25 watt, 5% tolerance		
R1, R2	10 kilohms (kΩ)	M10K
R3	1 kilohm (kΩ)	M1K0
RV1	100 kilohm (kΩ) min. preset (horizontal mounting)	UH06G
Capacitors		
C1	22 nanofarads (nF) ceramic, 25 V	WX78K
C2	47 nanofarads (nF) ceramic, 25 V	RA47B

Integrated circuit
   IC1   NE555   QH66W
Additional items
   8-pin DIL socket   BL17T
   Jack plug and socket 2.5 mm   HF76H, HF78K
   Jack plug and socket 3.5 mm   HF80B, HF82D
   Battery type PP3
   Battery connector   HF28F
   Plastic box with lid 114 × 76 × 38 mm   LH14Q
   Veroboard 0.1 inch with copper strips
      18 holes × 10 strips   JP46A
   Terminal strip (two section)   FE78K
   Crystal earpiece   LB25C
   Morse key

27 The Colt 80 m receiver – Part 4

Introduction

In Parts 1 to 3 of this series, the design of an 80 metre direct-conversion receiver has been described. In this final part, we are going to change the circuit to operate as a superheterodyne receiver, or *superhet*. Most radio receivers are superhets, and they change the incoming signal to another frequency, known as the *intermediate frequency*, or IF, before producing an audio signal. The use of a superhet in a good receiver is mandated by the requirements for good *sensitivity* and *selectivity*.

Sensitivity and selectivity

Figure 1 shows the block diagram of the receiver. If you look closely, you will see that it is two similar circuits, one after the other. The incoming

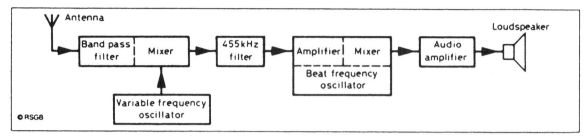

Figure 1 This is how the different stages of the Colt go together to make a superhet receiver

signal is filtered and fed to a mixer where it is combined (mixed) with the signal from the variable-frequency oscillator (VFO). This oscillator operates at a frequency which is 455 kHz higher than the incoming signal from the aerial, and the mixer output is therefore at a frequency of 455 kHz. If you are not sure about this, please refer to the section 'The direct conversion process' in Part 3 of this project. This new frequency is called the *intermediate frequency*, or IF. This frequency doesn't change; the tuning is accomplished by the VFO, and the mixer output is *always* at 455 kHz. The extraction of the audio signal from the IF signal is identical with the direct conversion process which is used in your existing receiver.

This may seem a long-winded way of doing things, but it has its advantages. A receiver must have a good *sensitivity*, or gain, so that it can receive very weak signals. In very general terms, it is easier to handle low-frequency signals than it is to handle high-frequency signals. We are changing our signal frequency from around 3.6 MHz down to 0.455 MHz (455 kHz), which is much lower and can be filtered and amplified relatively easily. A second advantage is that it is easier to provide gain at a fixed frequency than at a variable frequency. Remember that the IF is fixed, and providing gain is, again, relatively simple.

Our receiver also needs good *selectivity*, the ability to separate (or select) one station from another very close to it in frequency. This requirement is significantly simplified by the fact that the IF is fixed, and a good filter in the IF circuits can do wonders for the rejection of adjacent-frequency stations! Several stages of IF amplification and filtering are possible in more adventurous designs.

The filtered signal at 455 kHz passes to another mixer which has an associated oscillator, usually called a *beat frequency oscillator* (BFO). When receiving CW (Morse) signals, the BFO is usually tuned about 1 kHz above or below the IF (i.e. at 454 or 456 kHz) to produce a 1 kHz beat note as the audio signal. This signal is then amplified and fed to a loudspeaker or headphones.

The circuit

Figure 2 shows the circuit of the IF section shown in the photograph on p. 94; this section is added to the existing circuit to make it a superhet. The existing mixer board (described in Part 3) is used as the first mixer. Between it and the audio amplifier is connected the new IF board.

The signal output from the first mixer (at 455 kHz IF) is fed into a *crystal filter*, the most expensive part in the whole receiver. It provides the selectivity which makes the Colt such a good receiver. Another NE602 mixer/oscillator chip follows the filter. The oscillator section is controlled by the tuned circuit in T1, the frequency of which can be altered by rotating the core inside the coil. Once set, it remains fixed.

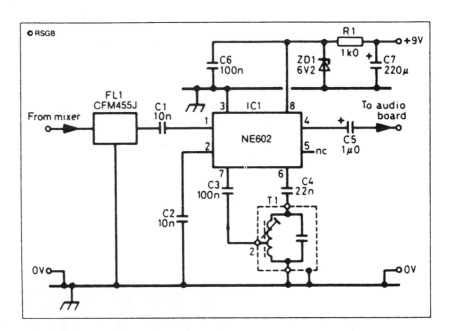

Figure 2 The IF board is connected between the mixer and audio amplifier

At this stage you should have built the first mixer and the audio amplifier, and proved that they *both* work by using the circuit as a direct-conversion receiver. When you have finished constructing and checking the IF board, you will need to add it to your existing circuits.

Adjusting and testing

There are three pairs of connections to the IF board – the 9 V supply leads, the input leads and the output leads. The IF board should be mounted on the metal baseplate, along with the other boards. **In making the following connections, make absolutely sure that the braid of each piece of coaxial** cable is soldered to the correct connection on each board. **The same applies to the polarity of the 9 V leads.** Disconnect the screened lead which presently goes from the first mixer board to the AF amplifier at the amplifier end. Connect it instead to the *input* of the IF board. Connect a new piece of screened cable from the IF output to the vacated AF input. Now connect the 9 V supply leads to the same points as the supply leads from the other boards. Check these newly made connections.

You should have confidence at this point that things should be right. After all, you *have* tested the direct-conversion process and you *know* it works. All that you are now testing is the IF board. This is the attraction of building a receiver in modules, working from the speaker backwards, and testing as you go!

The VFO and BFO need to be correctly adjusted. Mesh the vanes of the tuning capacitor and, using the same frequency measurement method as you did originally, set the VFO frequency to 3.955 MHz (which is 3.500 MHz + 0.455 MHz, if you hadn't guessed!). The BFO can be set using a frequency counter, but it is just as good to set it by listening to SSB or CW signals. A high-pitched hissing sound should be heard in the speaker. As you rotate the core, the pitch should reduce, go through a minimum, then increase again. Set the core at the minimum pitch position. You may want to readjust the two cores a little as your listening skills improve but, once you are happy, they will never need to be altered again!

In conclusion . . .

You should now have built a superhet receiver capable of excellent results. It uses the same type of circuit as that found in far more advanced receivers.

The superhet is far more sensitive than the direct-conversion type, and can weed out those elusive DX stations. You may have found that a station will appear at two places on the dial of the direct-conversion receiver; you will have no such problem with the superhet.

Parts list

Resistor
R1	1 kilohm (kΩ), 0.25 watt, 5% tolerance

Capacitors
C1, C2	10 nanofarads (nF) min. ceramic
C3, C6	100 nanofarads (nF) min. ceramic
C4	22 nanofarads (nF) min. ceramic
C5	1 microfarad (μF) 16 V electrolytic
C7	220 microfarad (μF) 16 V electrolytic

Integrated circuit
IC1	NE602

Additional items
ZD1	Zener, 6.2 V 0.5 watt
FL1	Crystal filter, Murata CFM455J
T1	Tuned inductor, Toko YHCS11100AC

---

28 A simple crystal set

Introduction

There are many designs of crystal set – they are all 'simple', and even the most seasoned radio amateur will build one of these every so often because it is something that never ceases to amaze! Although using modern components, the design is a period piece!

Design and construction

The circuit diagram of **Figure 1** shows the simplicity of the circuit. The prototype was constructed on a wooden baseboard with an aluminium front panel, although this could be made of wood if you prefer.

Winding the coil always causes the most groans but when finished, gives the most satisfaction. The construction of the coil is detailed in **Figure 2**. A discarded toilet roll centre is an ideal *former* on which to wind the coil. The coil will take up about 7 cm of the length of the tube, so assuming the tube is about 11 cm long, the ends of the coil will be about 2 cm in from each end of the tube.

The aerial coil, the *primary winding* of this radio-frequency (RF) transformer, is made from about 30 turns of 26 SWG enamelled copper wire, wound on the matchstick ribs. These ribs should be about 2 cm long, and are glued on **top** of the end of the secondary coil *after* the secondary coil has been wound. The *secondary winding* has 140 turns with *taps* every 10 turns for 70 turns. This coil covers most of the length of the cardboard tube. To make your coil-winding easier, here are some tips for completing the coil and still having some hair left at the end!

- Support the reel of wire on a dowel rod or pencil, clamped in a vice or held rigid in a vertical position by some other means. Don't leave it to trail around on the floor.
- Have a small piece of sandpaper handy to remove the enamel from the copper wire, starting with the end coming from the reel. Remove about 1 cm only, until the shiny copper is visible all round the end of the wire.

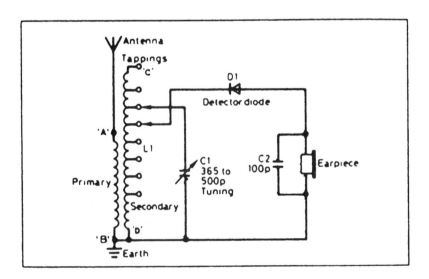

Figure 1 Circuit diagram of the crystal set

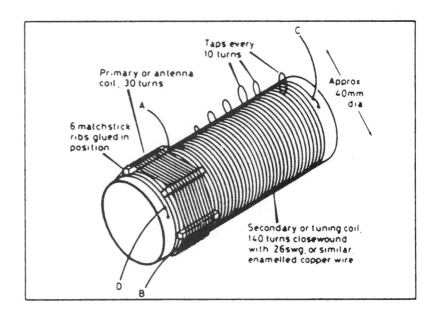

Figure 2 The coil is mounted on a cardboard tube as shown. Exact size is not too important. Try different taps for best results

- This part is the most difficult of all. Seasoned coil winders may have their own favourite ways of doing this, but for first-timers, play it safe and do it this way! Have a second bobbin (or a piece of stout card) ready, on which you will need to wind the wire loosely while you prepare the taps. Then, starting from the free end of the wire, measure 12.5 cm and, at this point, remove about 2 cm of the enamel with the sandpaper. Bend the wire firmly back on itself in the middle of the bared area, and *tin* (cover with solder) the exposed copper. This is your first tap. Then repeat this six more times at 12.5 cm intervals, thus giving you seven taps; wrap the wire as you go, on to the second bobbin or piece of card. After the seven taps, you are half-way down your coil so, to begin winding it in earnest, wind the wire carefully from the second bobbin or card **back** on to the original reel of wire.
- Looking at Figure 2, make two small holes about 3 mm apart, about 2 cm down from the top of the tube, where wire C will be entering. Poke the end of the wire into one hole, then bring it out again through the other; leave about 10 cm of wire on the free end. Loop this wire in and out once more, thus anchoring the wire firmly.
- Then, begin winding; keep the wire tight and make adjacent turns touch neatly. As you pass each tap, make sure it sticks outwards while avoiding flexing the wire too much; the wire is inherently weak at each tap.
- When you reach the last 10 cm of wire, stop. Put two more holes, like the first, beside your stopping point, and anchor the end, D, of the coil in the same way as you did with end C. The secondary winding is complete. Take a break *after* you have completed the next step!

● Prepare the six matchsticks for supporting the primary coil and glue these every 60 degrees around the end of the coil at D. Remove the enamel from the free end of the new copper wire on your reel and, using your favourite quick-drying cement, put a drop over the wire on one of the matchsticks, leaving 10 cm free as before. If you are using 26 SWG enamelled copper wire, its diameter is 0.46 mm, so 30 turns should occupy 14–15 mm of the 20 mm matchstick length. This will allow you to position the glued point such that the coil will be roughly in the centre of the matchsticks. When the glue has set, wind the 30 turns and anchor this end of the coil in the same way. Then remove the enamel at the end of the 10 cm lead.

Assembly

The coil former can be screwed to the baseboard at each end, *before* the front panel is screwed to the baseboard! As the photograph on the next page shows, the tuning capacitor is mounted in the centre of the front panel. Five solder tags (or drawing pins as a last resort) should be screwed into the baseboard. The two to the left of the photograph are the connections to the earphones, the two along the rear edge have the diode, D1, soldered to them. One end of the diode (and, for once, you can connect the diode either way round!) is connected to one earphone tag. The other end has a flying lead of about 10 cm attached to it and terminated in a crocodile clip which is used to connect to one of the taps on the coil. The fifth tag secures the end, A, of the primary coil, to which you will attach the aerial. The earth tag on the tuning capacitor, C1, serves as an anchor point for the ground connections to B and D and to the other earphone tag.

Use

The longest piece of wire you have available to use as an aerial should be connected to the aerial tag just mentioned. If you can connect the earth tag of the tuning capacitor to a good electrical earth, this will help also. You should be able to hear something as you turn the tuning knob. Try adjusting the tapping connections on the coil – change only one at a time, or you will never find the optimum positions. You now have a fully operational crystal set!

Simple it may be, but this circuit illustrates some important principles which are used even in the most expensive receivers. Firstly, the coil, L1, and the capacitor, C1, form a *tuned circuit* which *resonates* at the frequency of the station you have tuned in. This *selects* the signal you want to hear. The *modulation* on this signal is removed by the *detector*, D1, and fed to the earphones, which act as *transducers*, turning the electrical energy into sound energy which you can hear. It doesn't need a battery or other power supply either!

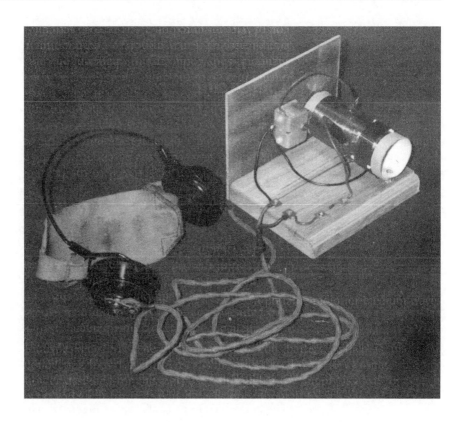

Parts list

Capacitors
- C1   Variable capacitor of between 200 and 500 picofarads (pF) maximum
- C2   100 picofarads (pF) min. ceramic

Additional items
- D1   Germanium, type OA47 or OA90

High impedance crystal earphone (not Walkman type)

Tuning knob

Wooden base, approx. 110 mm square

Aluminium (or wood) front panel, approx. 80×100 mm

Cardboard tube (toilet roll centre), about 110 mm long, 40 mm diameter

Reel of 26 SWG enamelled copper wire for both windings of L1

29 A crystal calibrator

Introduction

A crystal calibrator is a device that produces oscillations of a precise frequency which are rich in *harmonics*. A harmonic is an integer multiple of the fundamental frequency; for example, if our fundamental frequency was 100 kHz (as it is in this circuit), there would be harmonics of this frequency at 200 kHz, 300 kHz, 400 kHz, and so on. These harmonics can be used, when picked up by any receiver, to calibrate that receiver, as the harmonics are quite accurate in frequency (see later for an assessment of accuracy). However, a gap of 100 kHz between harmonics is rather wide for most purposes, so we reduce this frequency to 25 kHz, so that the harmonics are then 25 kHz apart, thus producing a much more useful set of *marker points*. A calibrator such as this is often called a *crystal marker*, producing these marker points from 25 kHz to beyond 30 MHz.

The circuit

The circuit diagram is shown in **Figure 1**. The circuit around TR1 is the fundamental oscillator, and its frequency is controlled by the quartz crystal, X1. Even crystal oscillators are not 100% accurate, and the small trimmer capacitor, TC1, is able to 'pull' the frequency to one which is nearer the

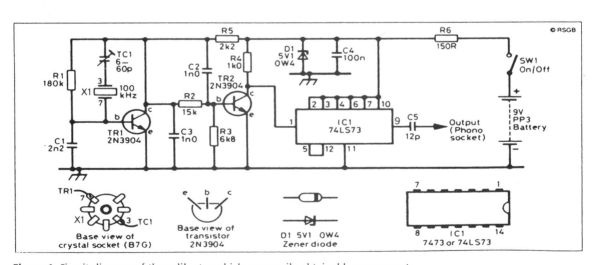

Figure 1 Circuit diagram of the calibrator which uses easily obtainable components

correct one. TR2 is a *buffer* stage, which isolates the oscillator from the rest of the circuit. It acts as a switch, and applies a good signal (switching between 0 and 5 V) to the input of IC1.

IC1 is an integrated circuit which can be connected to do several things. It is connected here to divide the incoming frequency by a factor of 4, producing on pin 9 an output frequency of 25 kHz.

The combination of R6, C4 and D1 produces a supply of 5 V for IC1; it would be damaged if the battery voltage were applied to it. You will no doubt be ready to assemble the circuit, so here is some information for you.

Construction

If possible, always build a circuit in individual stages, which you can test as you go along. It is not always easy in small projects like this, but even the crystal calibrator can be split into two for construction and testing.

It is an ideal project for assembly on Veroboard of the copper-strip type; the prototype layout is shown in **Figure 2**, on a board measuring 11 holes by 24 strips. First of all, remove the copper strip at the locations shown, using a 3 mm drill rotated between the fingers. Hold the board up to the light to ensure that there are no pieces of copper swarf bridging adjacent strips, and that the copper is completely removed where it should be! Assemble the circuit from left to right, but do not wire up anything around IC1. Be aware that the diagram shows the board *from the component side*. Although you

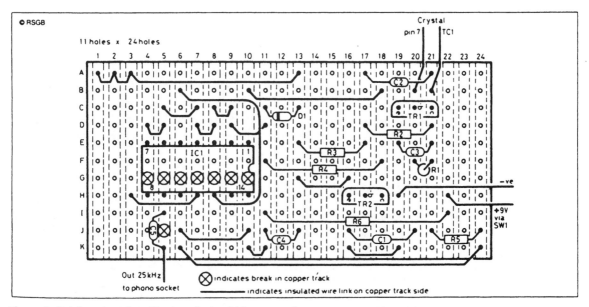

Figure 2 Stripboard, such as Veroboard, should be cut as shown to approx. 62 × 28 mm

can use virtually any kind of crystal, you will need to mount it firmly. If you use a small one, it can be soldered directly in position on the board. With one in a valve envelope, you will need a B7G valveholder. Be prepared to mount the circuit in a **metal** container; the prototype used an (empty) tin of tuna! The valveholder mounts well on any metal case.

Disconnect the aerial of your receiver, and replace it with about 30 cm of wire laid near your oscillator circuit. Connect the battery and switch on. Tune the receiver around until you can hear a whistle on SSB. Rotate the tuning knob to reduce the frequency of the whistle and, as the whistle becomes too low to hear, you have reached one of the calibration markers, and your frequency will be an integer multiple of 100 kHz. If you are already using a calibrated receiver, you will be able to verify this. Going up or down in frequency should locate another marker 100 kHz away, and so on right through the receiver's tuning range. If the signals coming from the circuit are very weak, switch off the calibrator and connect another 30 cm piece of insulated wire to the collector of TR2, and lay it close to (but *not* touching) the wire from the aerial connector. Switch on an try again; you should not now have a problem! If you have access to a multimeter, check that the voltage at the collector of TR1 is very close to 5 V. If it is close to 9 V, you have connected the diode, D1, the wrong way round!

Having verified that the oscillator is working, you can now wire up the integrated circuit socket, being careful to put the notched end of the socket pointing towards R4, the collector resistor for TR2.

Check your connections around IC1, and when you are satisfied that they are correct, line up IC1 with its socket, making sure that the notched ends are together, and press down *gently* to insert the IC into its socket. Insert the 30 cm piece of insulated wire into the output socket, connect the battery and switch on. You should still hear whistles in your receiver, but now they should be 25 kHz apart, rather than 100 kHz apart.

All that now remains to be done is to mount the circuit rigidly inside whatever casing your have chosen. Make sure that none of the connections under the board touch the metal case, and secure the valveholder, on/off switch and output connector to the case. You now have a completed crystal calibrator.

Calibration

The simplest way to calibrate your circuit is with a frequency counter. Most clubs will have one of these and, if not, will know someone who has! Connect it to the collector of TR2, where the frequency should be 100 kHz. Do *not* connect it to any part of the circuit around TR1, or you may alter the frequency you are trying to measure! If the frequency is not exactly 100.000 kHz, rotate TC1 until it is (or is as close as you can get it). Now your calibrator is as accurate as the counter with which you have calibrated it.

Accuracy

Despite your best efforts at calibration, by whatever means, your crystal will *never* have a constant frequency. Such a thing is a scientific impossibility. It is usual to express the accuracy of a crystal in parts-per-million (ppm), and it is governed by many things, principally its temperature. You will be *very* fortunate if your circuit maintains an accuracy better than about ±10 ppm. Expressed in figures, it means that the true frequency can be anywhere between 99.999 kHz and 100.001 kHz, i.e. within 1 Hz of the correct frequency.

Although this may seem more than adequate, it is as well to remember that the accuracy degrades as the frequency increases. At 1 MHz the error will be ±10 Hz and at 10 MHz it will be ±100 Hz. At 30 MHz it will be 300 Hz. Even so, this should be acceptable for most non-critical applications.

Parts list

Resistors: all 0.25 watt, 5% tolerance		Maplin code
R1	180 kilohms (kΩ)	M180K
R2	15 kilohms (kΩ)	M15K
R3	6.8 kilohms (kΩ)	M6K8
R4	1 kilohm (kΩ)	M1K0
R5	2 kilohms (kΩ)	M2K0
R6	150 ohms (Ω)	M150R

Capacitors		
C1, C2, C3	1 nanofarad (nF) or 1000 picofarads (pF)	WX68Y
C4	100 nanofarads or 0.1 microfarad (μF)	YR75S
C5	12 picofarads (pF)	WX45Y
TC1	60 picofarads (pF) trimmer	WL72P

Semiconductors		
TR1, TR2	2N3904 npn	QR28F
IC1	7473 or 74LS73N dual JK flipflop	YF30H
D1	5.1V 500 mW Zener	QH07H

Additional items	
100 kHz crystal	
B7G valveholder (found at most rallies)	
14-pin DIL IC socket	BL18U
Phono socket	YW06G
PP3 battery connector	HF28F
PP3 battery	
On/off switch	FH00A
Veroboard	JP47B
Metal case as required	

30 A simple short-wave receiver – Part 1

Introduction

No apologies for yet another design of short-wave radio. This is the beauty of our hobby – there is always something new to try. Not all circuits operate in the same way; not all circuits work equally well; sometimes a simple design suits the operator better than a complicated design. So here is another one for your consideration; it's a good project for a novice, *and* a good one-evening project for someone more experienced.

A basic description

Figure 1 gives the block diagram of the receiver, which employs a *regenerative detector*, one of the earliest techniques by which excellent *selectivity* (the ability to separate two stations very close in frequency) could be combined with good *sensitivity* (the ability to pick up very weak stations) using a very simple circuit. The receiver falls into the category known as *tuned radio frequency* (TRF), meaning that the whole circuit (prior to the extraction of the modulation from the carrier) operates at the incoming radio frequency. In other words, it is *not* a superhet.

As you can see, the regenerative detector has what is called a *feedback loop*, which feeds a small amount of the output signal back to the input. You have heard the effects of feedback with a public address system, when the microphone gain is too great – everything becomes very loud and then bursts into a deafening squeal! This is exactly what the feedback does here, except that it is carefully controlled, thus providing both gain and selectivity. The resulting audio is then amplified for use with headphones.

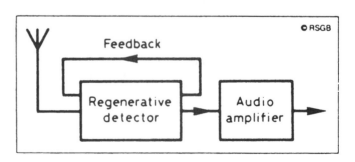

Figure 1 Just two simple stages make up the circuit

The circuit

The circuit of **Figure 2** shows the complete system. TR1 is an untuned *field-effect transistor* (FET) stage, and is used to match the aerial to the next stage. Occasionally, a regenerative detector produces unwanted signals, and TR1 also prevents them from reaching the aerial and being transmitted! The smaller of the two windings on T1 (the *primary* winding) will match a low-impedance aerial, the capacitor input matching a high-impedance aerial.

Ignore TR2 for the moment – the next stage in the signal path is TR3, the detector, using another FET. The tuned circuit is formed by T2, VC1 and VC2 (remember TR1 is untuned). The primary winding on T2 taking the output from TR1. The reason for having two variable tuning capacitors, one large, the other small, is that the large capacitor is the main tuning capacitor, while the small one is used as a *bandspread* control (for very fine tuning). The tuned signals are then detected (converted to audio frequencies) by TR3.

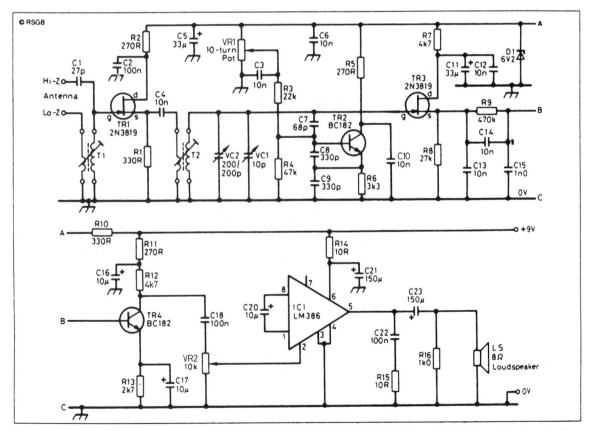

Figure 2 The circuit diagram showing the 10-turn potentiometer VR1. This is the regeneration control for the circuit

TR2 is a *Q-multiplier* stage. It is a Colpitts-type oscillator which uses C8 and C9 to give feedback and produce the oscillation. C7 couples the tuned circuit (and the input to TR3) to TR2. The tuned circuit thus controls the frequency of oscillation of TR2 and the signal passed on to TR3.

The secret of the ease of operation of this receiver is VR2, a 10-turn potentiometer (or *helipot*). The resistance wire is wound in the form of a helix, giving a much greater wire length than in a normal potentiometer, and the shaft must be turned ten times to cover the whole length. Helipots are very useful when very fine adjustments have to be made. Here, VR2 sets the regeneration (or *reaction*) level, depending on the type of signal you are receiving, as will be discussed in Part 2.

TR4 provides the first stage of audio amplification and, after the volume control, VR2, the audio amplifier integrated circuit, IC1, will drive a small loudspeaker or headphones.

In Part 2, the construction will be discussed, together with the choice of aerial, the parts list, and advice on using the receiver.

31 A fruit-powered medium-wave radio

Introduction

This is a one-evening project that will result in a working medium-wave (MW) radio, and will also teach you a little about the way electricity can be generated from the right metals and a little (safe) acid. All you need are three lemons or other citrus fruit, three pieces of copper and three pieces of zinc (or galvanised metal) for your power supply.

Construction

Figure 1 shows the circuit and **Figure 2** its layout on a simple 'plug-in' prototype board. The six pieces of metal are connected as shown, to wire the three lemons in series; use ordinary wire between each lemon and the next. If you have a meter to measure the total voltage, it should be about 1.8 V. Use a standard ferrite rod, and wind on it about 40 turns of single-conductor PVC-insulated wire.

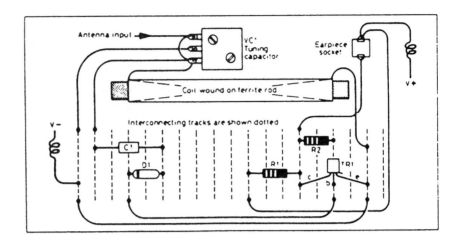

Figure 1 Most parts are plugged into the board as shown – soldering is not required

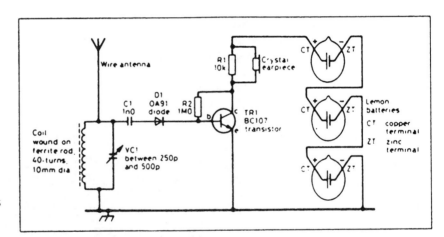

Figure 2 Three lemons power the radio, which gives good results for such a simple circuit

Wire up the circuit on the board as illustrated in Figure 2. Soldering is not required with this type of board – just plug in the components and the wires. Only one transistor is needed. The tuning capacitor, VC1, selects the station you want to hear, and D1 helps to remove the carrier from the RF signal. The resulting audio signal is fed to TR1, a small transistor audio amplifier, which makes the signal big enough to drive a crystal earpiece comfortably. Walkman-type earphones will not work, so invest in a crystal earpiece which you can use in several other projects, too! If you use a smaller capacitor than that specified for VC1, you will need more turns on the aerial coil.

Tests on the prototype indicated that the radio will run for about a week on three lemons!

Parts list

Resistors: all 0.25 watt, 5% tolerance
  R1   1 megohm (MΩ)
  R2   10 kilohm (kΩ)

Capacitors
  C1   1 nanofarad (nF) min. ceramic
  VC1   250–500 picofarad (pF) variable

Semiconductors
  TR1   BC107 npn (or BC108, BC109C)
  D1   OA90, OA91 germanium (not silicon)

Coil
  L1   2 metres of single-conductor insulated wire on
            a standard ferrite rod

Additional items
  Plug-in prototype board, e.g. Maplin YR84F
  Wire aerial at least 3 m long
  Crystal earpiece
  Three juicy fruits

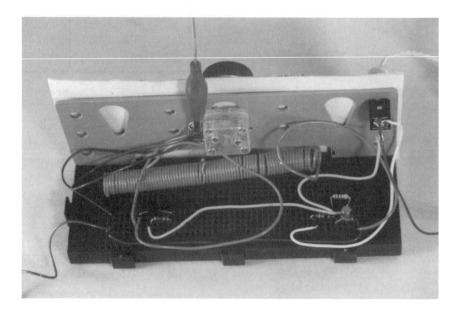

32 A capacitance bridge

Introduction

It is always valuable to have a means of measuring capacitors amongst your test gear. Universal LCR bridges (i.e. systems that can measure inductance, capacitance and resistance) can be found at most rallies, and you can work your way up to owning one of these. In the meantime, a piece of home-made equipment gives you experience as well as resulting in a useful measuring instrument.

The circuit described here is called a *capacitance bridge*, because it balances the effects of one resistor/capacitor pair against another; if one capacitor has an unknown value, then the other can be calculated. The basic bridge circuit is shown in **Figure 1**. To avoid having calculations to perform, this instrument will be calibrated by using capacitors of known values. The bridge is a useful way of performing measurements, because a knob is turned until there is a null in the signal from an earpiece or loudspeaker. The ear is very precise in being able to perceive nulls, which makes the bridge easy to use and reasonably accurate. At the null, the bridge is said to be *balanced*.

How does it work?

Figure 2 is the circuit diagram for this capacitance bridge. Transistors TR1 and TR2 form an oscillator. This is the audio oscillator shown in Figure 1, and produces an alternating voltage which is fed to the bridge. RV1 (in the collector lead of TR2) replaces both R1 and R2 in Figure 1 – that part of

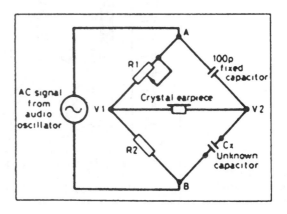

Figure 1 Simplified circuit of a capacitance bridge. R1 is adjusted for minimum sound

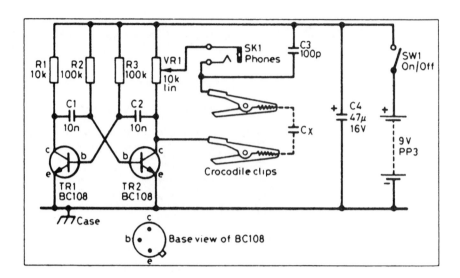

Figure 2 Transistors TR1 and TR2 give an audio signal which is adjusted by variable resistor RV1

RV1 *above* the wiper represents R1, while that part *below* the wiper represents R2. The voltage on one side of the earpiece is determined by the ratio of these values, and is adjusted by rotating RV1. The voltage on the other side of the earpiece is determined by the ratio of C3 to Cx, where Cx is the unknown-valued capacitor. When these two voltages are the same, the bridge is balanced and there is no current through the earpiece.

Figure 3 shows the layout of the components on the matrix board of the prototype. It measures 10 holes by 14 holes, and is of the plain type, i.e. no copper strips. All earth connections are taken to a single solder tag. When

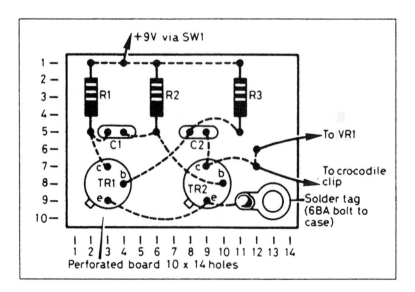

Figure 3 Components are soldered together on a small prototype board

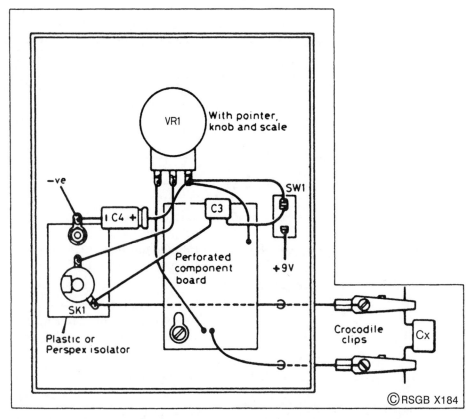

Figure 4 Layout of the parts inside an aluminium box. The earpiece socket is insulated from the case

mounting the board inside a metal box, a long screw is used with extra nuts, to earth the board to the case while providing a stand-off, thus preventing any unwanted short-circuits between the board connections and the case. Be aware that a *crystal* earpiece must be used; low-impedance headphones of the Walkman type are not satisfactory. Solder two flying leads as shown in **Figure 4**, about 15 cm long, terminated in small crocodile clips for attaching to the unknown capacitor.

Calibration

After checking the circuit carefully, attach the battery and switch on; a buzzing noise should be heard in the earpiece. This is the first sign that everything should be OK. If there is no buzz, switch off and recheck the connections.

You will now need a range of close-tolerance (1%) silver-mica capacitors covering the range 10 to 1000 picofarads (pF). Arrange the capacitors in ascending order and connect them to your bridge in sequence. After having prepared a neat piece of card or paper mounted behind the knob on the front panel, mark the dial at the positions of the nulls for all the capacitors. You have now calibrated your capacitance bridge. If you are more likely to want to measure larger capacitors, replace C3 by a 1 nanofarad (nF) capacitor, and the bridge will measure up to 10 nF approximately.

Parts list

Resistors: all 0.25 watt, 5% tolerance
R1	10 kilohms (kΩ)
R2, R3	100 kilohms (kΩ)
RV1	10 kilohms (kΩ) linear

Capacitors
C1, C2	10 nanofarads (nF) ceramic
C3	100 picofarads (pF) silver mica or polystyrene
C4	47 microfarads (μF) 16 V electrolytic

Semiconductors
TR1, TR2	BC108 npn

Additional items
SW1	SPST on/off switch
	Battery connector, PP3 type
	Earpiece, crystal type
SK1	3.5 mm jack socket for earpiece
	Battery, 9 V PP3
	Matrix board, approx. 10 holes by 14 holes
	Aluminium case, approx. $10 \times 8 \times 5$ cm

33 A simple short-wave receiver – Part 2

Introduction

In Part 1, the design of this receiver was discussed in some detail. Now we are going to put it all together and see how it works. The receiver is laid out on a printed-circuit board (PCB) or on a matrix board.

Construction

The layout of the components is shown in **Figure 1**. Identify each part separately, insert it into the holes in the board and solder carefully. Long leads may be cropped before or after soldering, depending on your skill and preference. All electrolytic capacitors, T1, T2 and IC1, must be connected correctly. The front-panel controls are connected to the PCB terminals shown in Figure 1.

The layout of the controls and the placing of the board inside the case are matters of personal preference. The size of the prototype front panel is shown in **Figure 2**. The prototype had a slow-motion drive fitted to VC2, the main tuning capacitor. This required the capacitor to be fitted on its own small panel. The bandspread control does not need any form of

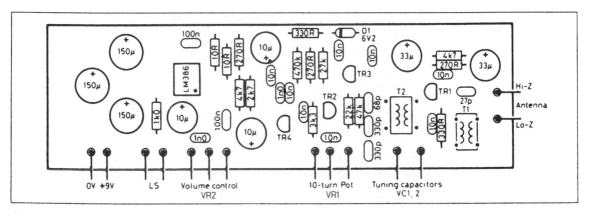

Figure 1 Layout of the receiver circuit board

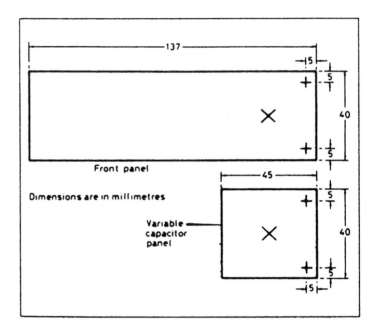

Front panel

Dimensions are in millimetres

Variable capacitor panel

Figure 2 The case can be metal or made from PCB material

slow-motion drive. A tuning dial may be fitted behind the tuning knob on VC2 if required. The volume control, VR2, has its own on/off switch for the battery, which can be mounted behind the back panel, on which are mounted the aerial input sockets and the headphone/speaker socket. Check your soldering carefully, together with all the wiring between the board, the potentiometers and variable capacitors. When you are confident that everything is perfect, connect the battery and switch on.

The first tests

If you have a good aerial and are using an aerial tuning unit (ATU), use the low-impedance input. Excellent results are possible, though, with about 3 metres of wire connected to the high-impedance input. Set the volume control to give a gentle hiss in the headphones or speaker. Advance the reaction control to give a definite hiss. As you tune in an AM station, the hiss will change to a whistle; back off the reaction until the oscillation *just* stops. The receiver is now correctly set for AM reception.

On the amateur bands, the stations will be SSB or CW, and the reaction needs to be set *just above* the oscillation point, and will need slight adjustment from time to time for different qualities of signal. Juggling with the volume, reaction and tuning is part of the pleasure of using regenerative receivers!

In action

Practice is needed for best results. The regenerative receiver is renowned for its versatility in being able to be set up *exactly* for all types and strengths of signal. The basic receiver tunes from 6.5 MHz to 11 MHz approximately; this includes two amateur bands at 7.0 MHz and 10.1 MHz and two broadcast bands.

Parts list

Resistors: all 0.25 watt, 5% tolerance

R1, R10	330 ohms (Ω)
R2, R5, R11	270 ohms (Ω)
R3	22 kilohms (kΩ)
R4	47 kilohms (kΩ)
R6	3.3 kilohms (kΩ)
R7, R12	4.7 kilohms (kΩ)
R8	27 kilohms (kΩ)
R9	470 kilohms (kΩ)
R13	2.7 kilohms (kΩ)
R14, R15	10 ohms (Ω)
R16	1 kilohm (kΩ)
VR1	10 kilohm (kΩ) linear 10-turn potentiometer
VR2	10 kilohm (kΩ) log potentiometer

Capacitors

C1	27 picofarads (pF)
C2, C18, C22	100 nanofarads (nF)
C3, C4, C6, C10, C13, C14	10 nanofarads (nF)
C5, C11	33 microfarads (μF) 16 V electrolytic
C7	68 picofarads (pF)
C8, C9	330 picofarads (pF)
C15	1 nanofarad (nF)
C16, C17, C20	10 microfarads (μF) 16 V electrolytic
C21, C23	150 microfarads (μF) 16 V electrolytic
VC1	10 picofarads (pF) variable
VC2	200 picofarads (pF) variable

Semiconductors

TR1, TR3	2N3819
TR2, TR4	BC182
IC1	LM386
D1	6.2 V 0.5 W Zener

Inductors
T1   3 turns primary, 15 turns secondary,
            wound on 2-hole ferrite bead, with 28 SWG wire
T2   2 turns primary, 17 turns secondary,
            wound on a T68-2 toroidal former, with 28 SWG
            wire

Additional items
  Printed-circuit board
  Battery connector
  PP3 battery
  8-pin DIL socket
  8 ohm speaker or headphones (Walkman type)
  Case to suit

34 A basic continuity tester

Introduction

This little device can be built on a plug-in breadboard, and is ideal for testing fuses and cables, as well as doubling as a signal source for testing amplifiers, etc. It can even be used for testing npn transistors by replacing either of the transistors in the circuit, and seeing if the circuit still works!

Simple and quick to build

Using a plug-in breadboard, this circuit is so simple it could almost be built when you need it, and then dismantled again! If you want the circuit to make a different sound, the components to change are R2, R3, C1 and C2. Always make sure that R2 = R3 and C1 = C2, or the sound may be excessively 'edgy' and lacking in volume.

Basically, the circuit is an oscillator which drives a little loudspeaker directly. The fuse or cable you are testing is connected between the crocodile clips. If there is a current path between the two clips, the current also flows through the circuit, thus operating the oscillator and producing a sound from the loudspeaker.

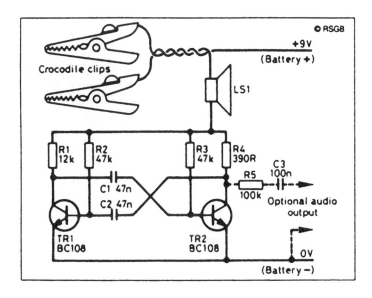

Figure 1 The circuit diagram of the continuity tester

If you want to use the circuit to produce a signal to test an amplifier, for instance, connect the croc clip which comes directly from the battery to the junction of R1, R2, R3 and R4. This supplies current to the circuit while, at the same time, bypassing the loudspeaker. Use the two components R5 and C3 between the oscillator and your amplifier. The loudspeaker is bypassed because you will want to listen to the output from your amplifier, *not* from the oscillator!

Remember, NEVER test any equipment which is still connected to the mains electricity supply. Avoid testing anything which is switched on and has its own power supply. You may damage both your tester **and** the circuit you are 'testing'!

Parts list.

Resistors: all 0.25 watt, 5% tolerance

R1	12 kilohms (kΩ)
R2, R3	47 kilohms (kΩ)
R4	390 ohms (Ω)
R5	100 kilohms (kΩ) optional

Capacitors

C1, C2	47 nanofarads (nF) ceramic
C3	100 nanofarads (nF), 0.1 microfarads (μF) ceramic optional

Semiconductors

TR1, TR2	BC108 npn

Additional items

LS1	Loudspeaker (8 Ω)
	2 crocodile clips
	Prototype circuit board
	PP3 battery connector
	PP3 battery

35 A charger for NiCad batteries

Introduction

The cost of replacing dry batteries can be alleviated to a great extent by the use of rechargeable batteries such as those containing nickel and cadmium (called *NiCads*). They are more expensive than the batteries they replace, but they can be charged hundreds of times and thus prove cheaper in the long run.

NiCads produce only 1.2 V per cell, compared with the 1.5 V of ordinary cells, so before you rush out and buy lots of these, please make sure that the equipment on which you plan to use them will work on the reduced voltage! For example, if you are using four cells to give you a 6 V supply, NiCads will give you only 4.8 V, which is quite a reduction. Using six cells to replace a 9 V battery will give you only 7.2 V. Not all equipment is happy with these reductions!

However, we all use them when we can, and they save substantial amounts of money. Here is the circuit of a charger to keep them in prime condition.

Charging NiCads – the ampère-hour

NiCads require charging at constant current, which means that connecting one across a normal power supply (constant *voltage*) is useless and can destroy it. They need pampering to the extent of needing a long charge (around 16 hours) at a rate dependent upon the *capacity* of the battery. By the capacity of a battery, we mean how much energy it can store. You will probably know that energy is measured in joules. For the purposes of storing energy in batteries, the joule is *not* the ideal unit, so we use one that is! This unit is the ampère-hour (Ah), and must be interpreted with some realism. For example, if the battery is rated at 2 Ah, it will deliver a current of 0.5 amp for 4 hours, or 0.25 amp for 8 hours. Provided the current is not too high, the product of the current (in amps) and the time for which it will flow (in hours) before the battery is flat will always be around 2 Ah. A workable 'rule of thumb' for calculating the charging current is that its value should be around one-tenth of the numerical value of the capacity; so, for our 2 Ah battery, a charging current of around 200 mA (2 ÷ 10 = 0.2 A or 200 mA) would be used.

Constant voltage to constant current

Many integrated circuit (IC) chips are available for use as voltage regulators, i.e. they supply a constant voltage. Most of these can be persuaded to become constant current supplies with one external resistor!

The voltage regulator IC usually has only three connections – 'input', 'output' and 'common'. It is designed (in the case of the LM7805) to produce a constant 5 V output between the 'output' and the 'common' connections, at currents of up to 1 A. If a resistor is connected between these, the IC will maintain 5 V across it. If you look at **Figure 1**, you will see the circuit performing this conversion.

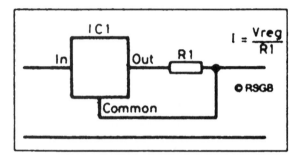

Figure 1 Voltage regulator arranged to produce a constant current

For the previous example, we derived a charging current of 200 mA, so we now need to calculate the value of resistor that will produce this current. Using the equation which is derived from Ohm's law:

$$I = \frac{V}{R}, \text{ from which } R = \frac{V}{I},$$

where R = R1, the resistance in ohms that we are calculating,
        V is the voltage across R1 (5 V), and
        I is the current flowing (200 mA).

So,

$$R1 = \frac{5}{0.2} = 25\,\Omega.$$

25 ohms is not a 'common' or 'preferred' resistor value, so we must choose the next *largest* value, which is 27 Ω. This reduces the current, but only slightly – it is now 185 mA. When calculating resistor values in power supply circuits, we must *always* check the power that they dissipate and make sure we specify and fit a suitable resistor.

Power (in watts) is the product of the voltage across and the current through a device, so in this case it is given by:

$$\text{Power} = V \times I = 5 \times 0.185 = 0.925\,\text{W}.$$

Rather than use a 1 watt resistor operating very near its limit, it is safer to use a 2 watt resistor operating well within its limits.

Looking again at Figure 1, we now have a constant-current source producing 185 mA, when R1 is a 27 Ω, 2 W resistor. For use with NiCads requiring charging currents other than 200 mA, you will need to repeat the two equations above, using a new value for I.

The full circuit and its assembly

This is shown in **Figure 2**, and can be broken down into two parts. The first is the AC to DC conversion produced by the mains transformer, T1, the bridge rectifier, BR1, and the smoothing capacitor, C1. The second is the constant-current section already discussed.

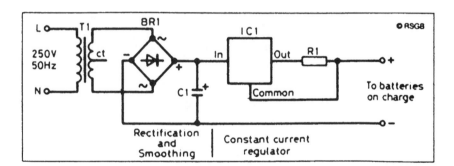

Figure 2 The complete circuit including mains components

The prototype was assembled on matrix board measuring 18 holes by 12 strips, although, as **Figure 3** shows, this is much larger than is strictly necessary. No strip cutting is needed, but make sure that IC1 and C1 are inserted correctly.

Warning! Before you attempt to wire up the transformer and the bridge rectifier, be aware that you will eventually be connecting the circuit to the mains supply, so there are three possibilities for you: (1) get a qualified friend to supervise your completion and testing of the circuit; (2) get your qualified friend to complete and test the circuit for you; (3) replace the transformer and bridge rectifier with a mains adapter.

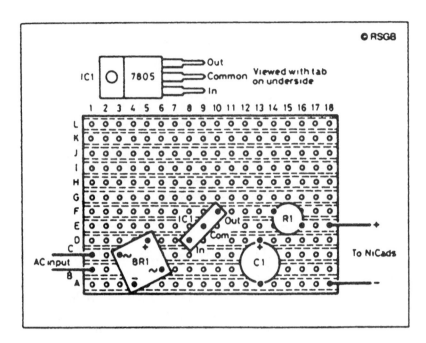

Figure 3 The charger can be built on Veroboard

If you decide to use the mains adapter, its output is connected directly across C1, because T1 and BR1 are now no longer needed. Make sure the polarity (positive and negative) is correct and that the adapter output is set to 12 V.

A quick test

If you have built the mains version, make sure all connections are correct, and that there are no soldered joints which will touch other parts of the circuit. The box must be securely closed before tests begin. The RSGB cannot be held responsible for damage to equipment or batteries! The version using the mains adapter need not be closed during tests.

Switch on. With nothing connected to the output, the unit should run cold. If this is not the case, switch off and recheck your circuit. If all is well, switch on again and connect a DC multimeter (on the current range) across the output. It should indicate only a slight difference from the calculated value of 185 mA. You can now charge your NiCads!

Other charging currents can be set by having different values for R1, perhaps selectable by a rotary switch. Remember to make sure that the values of both resistance *and* power dissipation are correct, and don't exceed the 250 mA rating of the transformer (or the 1 amp rating of the IC if you are using a bigger transformer).

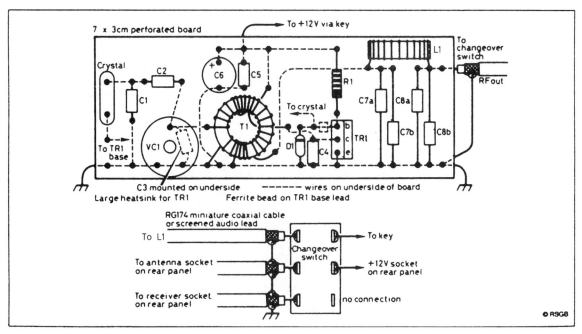

Figure 2 Component layout is straightforward on a 'matrix board' (it has holes but no copper strips). The dotted lines are the connecting wires. The lower part shows the connections for the transmit/receive switch

The output filter, which comprises C7, C8 and L1, is a *low-pass* filter, which helps reduce any *harmonics* present in your signal. Harmonics are integral multiples of your transmitter frequency, so if you are transmitting on a frequency, *f*, harmonics will be present at frequencies 2*f*, 3*f*, 4*f*, . . . and so on. L2 is another inductance using 22 SWG enamelled copper on a ferrite toroid. The changeover switch is external to the transmitter board, and is used to switch your aerial between the transmitter and the receiver; its wiring is shown in Figure 2.

Use a dummy load

A dummy load enables you to test your circuit without actually transmitting a signal. If you haven't such a thing already, it is easy to construct one to use with this transmitter. Don't use it for transmitters of more than 2 watts output, though. Use two 100 ohm, 1 watt resistors, connected in parallel across the end of a short piece of coaxial cable, terminated in a BNC, PL259 or N-type free plug, as shown in **Figure 3**. Plug this into the aerial socket on your transmitter, plug in your crystal and connect the transmitter to a 12 V supply. Have another receiver switched on and tuned to the crystal frequency. Although the radiation from your dummy load is minimal, it will

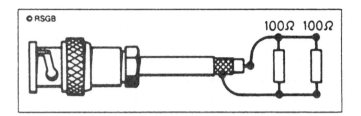

Figure 3 A simple 2 W dummy load can be made from two 100 Ω (ohm) resistors in parallel. The plug should match the socket on your transmitter

be enough to be picked up by a receiver in the same room. Send dashes with the Morse key, and adjust VC1 until the received note is *clean*. It should not sound rough, or have a *chirp* (change its frequency during a dash or dot). Avoid tuning for maximum power; this is seldom the correct setting!

You will need to put your completed transmitter in a metal box, using sockets for the power supply, aerial, receiver and Morse key. The sockets can be chosen to match your existing equipment.

Figures 4–7 are taken from the RSGB book *Practical Antennas for Novices*, and may give you some ideas on the type of aerial to be used with your transmitter.

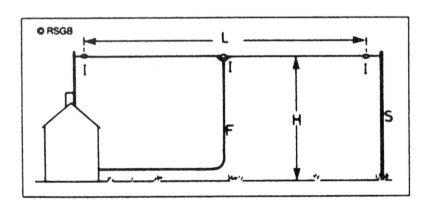

Figure 4 A simple dipole can be very effective. For the 3.5 MHz band, length L is 40 metres and height H should be as large as possible. The far support S can be a tree, pole or building. Insulators I may be home made from strong plastic and the feeder F should be 50 Ω (ohm) coax cable

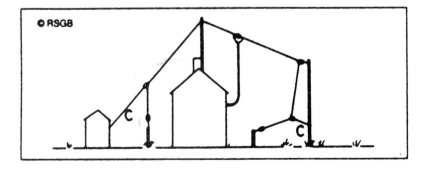

Figure 5 Your signal is radiated mostly from the centre of the dipole so the ends can droop or even be bent but the length may need shortening by a few centimetres because the ground and the bends will detune the dipole. Cords C are best made from strong plastic rope from a sailing or camping shop

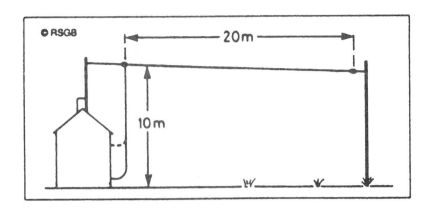

Figure 6 An 'inverted-L' takes up less space than a dipole and doesn't need coax cable. Like the dipole, the end can droop or be bent to save space as in this case most of the radiation comes from the area around the top of the vertical part

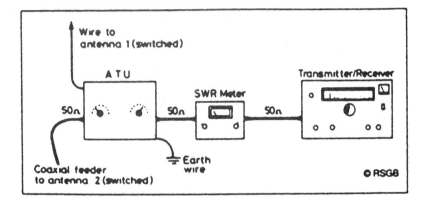

Figure 7 Almost any length of wire more than 10 m or so long will work (though it will work better the longer and higher it is) but an Aerial System Tuning Unit (ASTU or ATU) will be needed

127

Parts list

Resistor
R1   12 kilohms (kΩ), 0.25 watt, 5% tolerance

Capacitors
C1	1000 picofarads (pF) polystyrene
C2	100 picofarads (pF) polystyrene
C7a, C8a	680 picofarads (pF) polystyrene
C7b, C8b	68 picofarads (pF) polystyrene
C3, C4	100 picofarads (pF) ceramic
C5	10 nanofarads (nF) polyester
C6	10 microfarads (μF) electrolytic 25 V
VC1	100 picofarads (pF) trimmer

Semiconductors
D1	1N4148
TR1	2SC2078 (see sources list)

Inductors
T1	38 turns 26 SWG enamelled copper on T–50–2 toroid, with two link windings of four turns
L1	21 turns 22 SWG enamelled copper on T–50–2 toroid

Additional items
Ferrite bead
Crystal (e.g. 3.579 MHz) and holder
Metal box
Socket for Morse key.
  **This must be totally isolated from the metal
of the box, as** *both* **connections can be at +12 V.**
Sockets for 12 V supply, aerial and receiver
Switch – DPDT
Heat sink for TR1
RG174 miniature coaxial cable for signal leads (see Figure 2)

Component sources

Special components
2SC2078 Cricklewood Electronics Ltd, 40 Cricklewood
         Broadway, London, NW2 3ET.

37 A solar-powered MW radio

Introduction

What could be more ecologically friendly than a radio powered by the sun's energy? This design is quite standard, and if you have built any of the other medium-wave radios in this series, then this one should present few problems.

The solar panel

The solar panel is to the solar cell as the battery is to the cell; in other words a solar panel is several solar cells connected in series. The solar panel quoted for this radio will generate about 9 V at a current of around 30 mA on a sunny day. The circuit will operate on a supply of around 2 V, so bright sunshine is *not* necessary for satisfactory operation. The volume will be less, of course.

The circuit

The radio uses the ZN415E integrated circuit (IC), connected as shown in **Figure 1**, the complete circuit diagram. The signal is tuned in by the combination of L1 and VC1. L1 is made by winding about 35 turns of 24 SWG enamelled copper wire on a ferrite rod. A standard ferrite rod of 10 cm length and 1 cm diameter is used.

Signals selected by the tuned circuit are passed to IC1, which amplifies the signals and removes the audio component, which is then amplified further by IC2 for driving a small loudspeaker. The removal of the audio component is the process we call *detection*. In addition to this, IC1 provides *automatic gain control* (AGC), which helps to keep the audio signal constant, even when the incoming RF signal may vary due to fading.

The prototype board

Veroboard (also known as matrix board or stripboard) is ideal for the construction of the radio. The layout is shown in **Figure 2**. The board size is 11 strips by 30 holes (**please note that there is no row 'I', so take care with your counting!**). Using a 3 mm (⅛ inch) twist drill rotated between

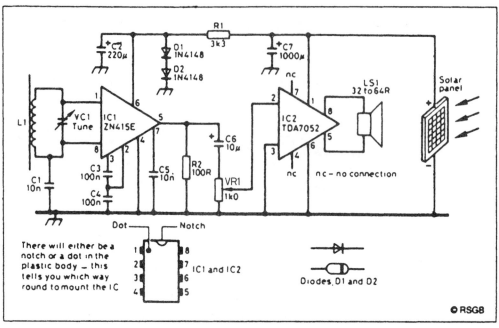

Figure 1 The radio uses just two integrated circuits (chips) and operates at any voltage from 2 to 9 V

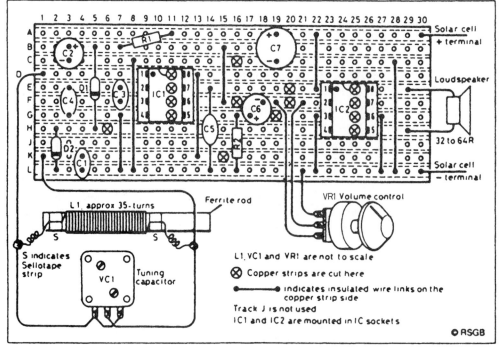

Figure 2 Follow the layout carefully, making sure that all connections are neatly soldered

your thumb and forefinger, break the strips at the points marked with a cross inside a circle. Hold the board up to the light to ensure that the tracks *have* been broken and that there is no copper swarf shorting tracks together.

Firstly, solder in the 8-pin DIL sockets for the ICs, making sure that the notches in the sockets are facing upwards, as shown on the diagram. Then, solder in the wire links, resistors, capacitors and diodes, making sure that the electrolytic capacitors (C2, C6 and C7) and the diodes (D1 and D2) are connected the correct way round. Use different colours of wire for the connections to the volume control, VR1, to avoid incorrect connections. Note the wiring of the tuning capacitor (VC1) shown in Figure 2; a two-section type is used, and both sections are wired in parallel to give twice the capacitance of a single section.

There is no on/off switch – just turn the volume down when you are finished using the set! The solar panel can be mounted parallel to the top of the case, or angled to receive the maximum energy from the sun, as shown in the photograph. You could have a battery available as a standby source to use the radio after dark; any battery of between 6 V and 9 V will do. Wire it with a simple changeover switch, so you can switch between solar and battery power. Ask a friend for help with this if you are not sure how to do it.

You may need to adjust the number of turns on L1 to get the best results, but it should be possible to receive at least five stations at good volume with your sun-powered radio!

Parts list

		Maplin code
Resistors: 0.25 watts, 5% tolerance		
R1	3.3 kilohms (kΩ)	M3K3
R2	100 ohms (Ω)	M100R
VR1	1 kilohm (kΩ) log	FW21X
Capacitors		
C1, C5	10 nanofarads (nF) ceramic	BX00A
C2	220 microfarads (μF) electrolytic (10 V)	FB60Q
C3, C4	100 nanofarads ceramic	YR75S
C6	10 microfarads (μF) electrolytic (10 V)	FB22Y
C7	1000 microfarads (μF) electrolytic (10 V)	FB81C
VC1	140–300 picofarads (pF)	FT78K
Semiconductors		
IC1	ZN415E (or ZN416E)	UR70M
IC2	TDA7052	UK79L
D1, D2	1N4148	QL80B
Solar panel	9 V at 50 mA	RK23A
Additional items		
LS1	32–64 ohm miniature loudspeaker	YT28F
	Ferrite rod	YG20W
	24 SWG enamelled copper wire	BL28F
	Plastic box, approx. 220 × 140 × 70 mm	YN39N
	Veroboard, cut to size	JP47B
	8-pin DIL sockets, 2 required	BL17T
	Knobs, 2 required	FD67X
	Material for speaker grille	
	Connecting wire	

38 A receiver for the 7 MHz amateur band

Introduction

Listening on the 40 metre band (from 7.0 to 7.1 MHz) can be very rewarding – it is a popular haunt for HF Special Event stations, and at night there are signals to be heard from all over Europe. This receiver is designed purely for the 40 m band, and is ideal for those who have built the simpler receivers and are looking for something a little more challenging. The more experienced constructor may prefer to build this on prototype board.

The circuit and its construction

Figure 1 shows the circuit diagram. The receiver will work well with headphones or loudspeaker. Walkman-type headphones and speakers are ideal for use here.

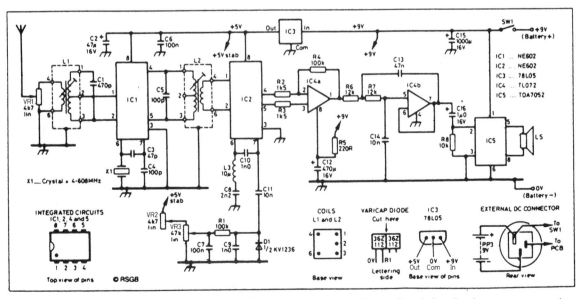

Figure 1 The receiver gives good performance on the 7 MHz amateur band as well as being simple to construct and align

133

Signals arriving at the aerial are coupled into IC1 via gain control VR1, which also functions as the on/off switch. Tuning is provided by varying the voltage on the *varactor diode* (or varicap), D1: VR3 is the main tuning control, and VR2 is the *bandspread* (fine tuning) control. The varactor diode is supplied as a dual device, which must be cut down the middle *carefully* with a sharp knife; with the lettering upwards, the ground lead (0 V) is on the left-hand side, as Figure 1 illustrates.

Solder in the IC sockets first, followed by the coils. After this come the links, resistors, capacitors and varactor diode. Ensure that IC3, the voltage regulator, is wired correctly, and check the polarity of the electrolytics. The crystal, X1, is very fragile, so take extra care with it. The wiring of the three controls is shown in **Figure 2**.

Before putting the ICs in their sockets, connect up the battery and check the following voltages with the negative voltmeter lead connected to the negative terminal of the battery:

Pin 8	IC1	5 V
Pin 8	IC2	5 V
Pin 8	IC4	9 V
Pin 1	IC5	9 V

When all these have been found to be correct, switch off and put the ICs carefully into their sockets. Use wire of different-coloured insulation to wire up the front-panel controls.

The case can be a small plastic box of size 22 cm by 15 cm by 8 cm, with three 10.5 mm holes drilled in the front and two 8 mm holes in the side for the aerial and earth connections. On one side are a 6 mm hole for the speaker socket and an 11 mm hole for the optional external power supply.

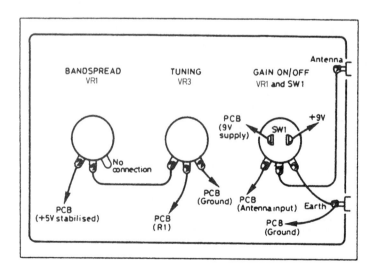

Figure 2 Rear view of the variable resistors. Check the connections carefully to make sure the wires are fitted correctly

Testing and tuning

The aerial for the receiver should be between 30 and 70 feet of wire, mounted as high as you can make it, away from trees and buildings if possible. Connect the battery and switch on. Adjust L1 and L2 for the best results. Tune slowly with VR3; you should find CW stations at the lower end of the band (anticlockwise) and SSB stations at the upper end (clockwise). You may find that it is easiest to make these adjustments *before* mounting the board in the case with double-sided sticky tape or pads. If you are planning to use an external DC supply, make sure it is a safety approved stabilised 9 V type, and **disconnect the battery before you use such a supply!**

If you suspect that the tuning doesn't quite cover the lower CW end of the band, try increasing C9 to 1200 pF. If it is the upper SSB end which is missing, decrease C9 to 820 pF.

It is always advisable to use an aerial tuning unit (ATU) between your aerial and the receiver. A suitable design of ATU is included as a project in this book.

Parts list

Resistors: all 0.25 watt, 5% tolerance

R1, R4	100 kilohms (kΩ)
R2, R3	1.5 kilohms (kΩ)
R5	220 ohms (Ω)
R6, R7	12 kilohms (kΩ)
R8	10 kilohms (kΩ)
VR1	4.7 kilohms (kΩ) linear, with SPST switch
VR2	4.7 kilohms (kΩ) linear
VR3	47 kilohms (kΩ) linear

Capacitors: all rated 16 V or more

C1	470 picofarads (pF) polystyrene 5%
C2	47 microfarads (μF) electrolytic
C3	47 picofarads (pF) polystyrene 5%
C4, C5	100 picofarads (pF) polystyrene 5%
C6, C7	100 nanofarads (nF) ceramic
C8	2.2 nanofarads (nF) polystyrene 5%
C9, C10	1 nanofarad (nF) polystyrene 5%
C11, C14	10 nanofarads (nF) ceramic
C12	470 microfarads (μF) electrolytic
C13	47 nanofarads (nF) ceramic
C15	1000 microfarads (μF) electrolytic
C16	1 microfarad (μF) electrolytic

Inductors
L1   Toko KANK3335R
L2   Toko KANK3333R
L3   10 μH 5%, e.g. Toko 283AS–100

Semiconductors
IC1, IC2   NE602 or NE602A
IC3   78L05 5 V, 100 mA
IC4   TL072
IC5   TDA7052

Additional items
D1   Toko KV1236 cut into two sections (one half used)
X1   4.608 MHz (available from Cirkit)
            3 × silver knobs, one with pointer
            Plastic case approx. 22 × 15 × 8 cm
            Speaker 8–32 Ω, or headphones
            4 × 8-pin DIL sockets for IC1, IC2, IC4, IC5
            2 × 4 mm sockets (red and black) for aerial and earth
            3.5 mm chassis-mounting jack socket for speaker
            DC power socket for external supply (if required)
            Prototype board

Kits

A complete kit is available from JAB Electronic Components.

39 Diodes for protection

Introduction

Many semiconductor devices can be destroyed in an instant if their supply is reversed. With the use of batteries as power sources, it is quite common for the battery to be connected 'the wrong way round', and the scope for damaging equipment is significant and very real. Diodes can protect equipment in several ways, and you may do worse than to consider one of these approaches to protect your next expensive project.

Choose wisely!

All the diode circuits given here are so simple as to invite calamity. The circuits are not foolproof but, with a little care, will work first time. Remember that a semiconductor diode has a forward voltage drop of between 0.5 V and 0.7 V, depending on its type and the current flowing. This will be mentioned later.

The first thing you need to do is to insert a good multimeter *in series* with the circuit you want to protect; set the range to Amps DC, and switch on. Check that the circuit works properly. Then, decrease the current range on the meter until a good reading is indicated. Make a note of this current, as it is the normal running current of your circuit.

To choose a diode, you must consult the catalogues or data sheets and find one where the quoted *maximum forward current* exceeds the current you have measured; preferably it should be at least twice your measured current. Secondly, the diode will have a *peak inverse voltage* (PIV); this is the maximum voltage it can withstand when the cathode is made positive with respect to the anode, i.e when it is *reverse-biased* and not conducting. This voltage must be greater than your battery voltage, again by a factor of about 2. Except in the case of the bridge rectifier (see later), these criteria will enable the selection of a suitable diode to be made easily.

The series diode

The simplest and most obvious way to protect equipment is to insert a diode in the positive supply lead, as shown in **Figure 1**, with the diode passing current *only* when the supply is of the correct polarity. Because of the 0.6 V that exists across the diode, your equipment will normally operate on a

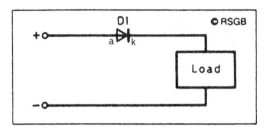

Figure 1 Series diode protection

slightly lower voltage. If you imagine reversing the supply to this circuit, you will see that the negative terminal of the battery is connected to the anode of the diode; the diode becomes reverse-biased and will not conduct any current, thus protecting the load. Your equipment will not operate when the battery connections are reversed.

The parallel diode

The circuit of **Figure 2** overcomes this voltage drop. It places the diode in parallel with the load (your equipment) but in a normally reverse-biased condition so that it draws no current when the battery is correctly connected. Reverse the battery connections, however, and a very large current will flow through the diode, thus blowing the fuse! For this technique to work successfully, the current drawn by the diode when the battery connections are reversed *must* be much greater than the maximum current drawn by the equipment, in order to blow the fuse. This is usually not a problem, however.

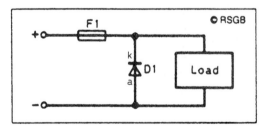

Figure 2 Parallel diode and fuse

The diode bridge

For sheer elegance, the circuit of **Figure 3** takes the biscuit! It uses four diodes connected in the form known as a *bridge rectifier*. Such rectifiers exist, and do not have to be made up from four discrete diodes.

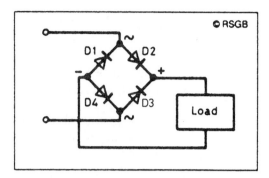

Figure 3 The diode bridge

Follow the current round the circuit from the supply, assuming initially that the top wire is positive and the bottom wire negative. It flows from the positive supply

(a) through D2
(b) through the load (top to bottom)
(c) through D4

and back to the negative of the supply.

Now assume that the bottom supply lead is positive and the top lead negative. The current flows from the positive supply

(a) through D3
(b) through the load (top to bottom)
(c) through D1

and back to the negative of the supply.

So, whichever way round the battery is connected, the current will *always* flow the same way through your equipment!

The circuit does have a drawback, however. Whichever way round the battery is connected, there are always **two** diodes conducting the current at any time. In the first case it is D2 and D4; in the second case, D1 and D3. This means that there is a total voltage drop of about 1.2 V. If your equipment can tolerate that reduction in voltage, then you will not have a problem.

Decoupling

Whenever the supply rail to a piece of equipment, or even to an individual stage of a circuit, is broken for the insertion of a device that will drop voltage, strange things can happen. This is because the supply for any circuit is assumed to have a low resistance to DC and a low *impedance* to AC. (Impedance is the AC equivalent of DC resistance.) These two are not the same, and the insertion of a diode or diodes is certain to make a big difference to them both. To overcome any irregularities in the operation of the circuit you are protecting in any of the ways described previously, the protected supply needs to be *decoupled* with a capacitor.

If the circuit handles audio frequencies only, placing an electrolytic capacitor directly across the load in all three circuits should solve the problem. The parallel circuit of Figure 2 is less at risk than the other circuits. The size of the capacitor will be determined by the current taken by the circuit, and may need to be chosen within the range 100 μF to 10 000 μF, with a working voltage greater than the supply voltage.

If the circuit is mainly handling RF currents, placing a capacitor of 0.01 μF across the load should prevent any problems. A second capacitor, also across the load, of between 10 μF and 100 μF may be needed. Again, the parallel circuit of Figure 2 is less at risk than the other two.

Don't be afraid to experiment, but confine your experimenting (at first) to small equipment and low currents, until you get a 'feel' for the technique.

40 An RF signal probe

Introduction

A radio-frequency (RF) diode probe is a simple device which, when used with a conventional multimeter, enables the measurement of RF voltages in a circuit. When constructed, this will be one of the most useful pieces of test equipment for the experimenter who revels in the construction of transmitters and receivers.

The circuit

Figure 1 shows the simple circuit diagram. It is *almost* the same as a common diode rectifier circuit, but with a simple change to make it more sensitive. The circuit is known as a *voltage-doubler*, and is often found in high-voltage supplies, with beefier capacitors and diodes, of course! Because we are dealing with high frequencies and smaller voltages, the diodes and capacitors can be physically very small. The diode circuit of D1 and D2 *rectifies* or *detects* the RF from the probe, and any remaining AC is removed (short-circuited to ground) by C2. This produces, at the output, a constant voltage proportional to the peak-to-peak RF voltage present at the input; the output voltage is fed to an ordinary multimeter (on a voltage range).

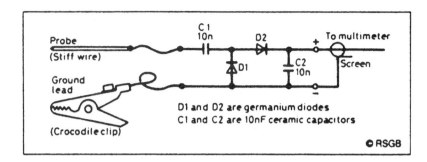

Figure 1 The diode probe can be made from four electronic components

Construction

Although the circuit layout is not critical, a description of the prototype is given here for your information. **Figure 2** has the details. The components are soldered to square pieces of copper-clad printed-circuit board (PCB) glued to a larger piece of PCB. The larger piece serves as a ground connection. Use a stiff copper wire as the probe, and an insulated flexible wire with a crocodile clip to connect the probe to the ground of the circuit under test.

Cut a piece of plain PCB, 30 mm by 45 mm and another of 15 mm by 45 mm. Cut the smaller piece into three measuring about 15 mm square. Stick the three small pieces to the larger piece, ensuring that there are small gaps between each, as Figure 2 shows. Solder the components in place. The connection to the multimeter should be thin coaxial cable or screened microphone cable. If you use unscreened cable, there may be RF pickup here which can lead to false readings.

Simple to use

Using the probe is simple. Connect it to the multimeter and set the meter to around 10 V DC – you may need to reduce this, depending upon the magnitude of the RF voltage you are trying to measure. Hold the probe by

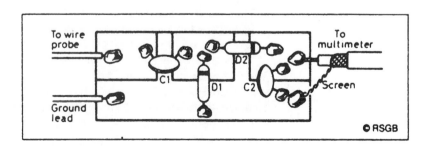

Figure 2 The components are mounted on a copper-clad PCB

its base, being careful *not* to touch any of the components. Connect the croc clip to a ground point on the circuit and touch the probe on the circuit point to be tested. If an RF signal is present, there should be a reading on the multimeter. That's all there is to it!

Parts list

Capacitors
  C1, C2   10 nanofarads (nF), or 0.01 microfarad (μF) ceramic

Semiconductors
  D1, D2   Any *germanium* signal diode, e.g. OA91 or AA119

Additional items
  Stiff copper wire
  Insulated flexible wire
  Copper-clad PCB
  Crocodile clip
  Coaxial or screened cable

Source

All components are available from Maplin.

41 An RF changeover circuit

Introduction

The simplest entry route to operating an HF station is to buy or build a receiver, and to add a simple CW (Morse) transmitter. In theory, this sounds so simple, yet in practice there is one significant hurdle to be overcome. How can they both share the same aerial? A manually operated changeover switch is the obvious solution, but it takes time to perform the switching operation, by which time someone else has squeezed in before you and contacted the DX station! This small circuit accomplishes the changeover automatically, as soon as the transmitter is keyed.

How it's done

The circuit is shown in **Figure 1**, and uses the signal from the transmitter to operate a relay. A relay is a switch which is operated electrically in the following way. A coil of wire with an iron core is used as an electromagnet. When current flows through the coil, it produces a magnetic field which, in turn, is used to pull a set of switch contacts. These contacts will be used to switch the aerial from transmitter to receiver and vice versa. The relay used here has a double-pole changeover (or DPDT, or DPCO) switch. If you are confused by the different types of switch, look at the basic descriptions elsewhere in this book. The *pole* of a switch is the part that doesn't move; in Figure 1, one pole is connected to L1 and the other to the aerial output. In a circuit diagram, the switch contacts are always shown in their *normal* state, i.e. when *no* current flows in the relay coil. The changeover switch is normally in the receive position, so we say the receive switch is *normally closed*, and the transmit switch is *normally open*. We choose to have the circuit in the receive position normally, because everyone spends a lot more time receiving than transmitting. The relay is energised only when you are transmitting.

There are three RF sockets – one for the transmitter, one for the aerial, and one for the receiver. When the transmitter is not in use, there is a direct connection (via the Rx normally closed contacts of the relay) between the aerial and the receiver. The transmitter (which is not keyed at this time) is connected to a 50 ohm dummy load, R1.

When the transmitter is activated by pressing the Morse key, an RF signal appears at the transmitter socket. The wire carrying the signal to the relay

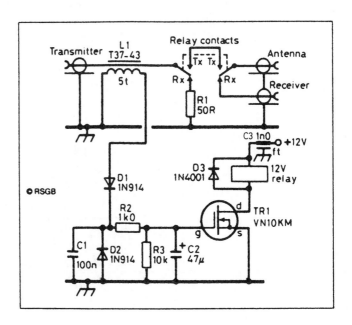

Figure 1 RF activated relay

143

Tx contacts passes through a toroid (a ferrite coil former, looking like a black Polo mint), which has a five-turn coil wrapped around it. This five-turn coil picks up the signal, which is then detected by D1 and D2, and converted to a steady voltage. C1 removes any remaining RF and C2 provides *hang*. Without it, the circuit would detect the gaps between every dot and dash being sent, and switch the aerial over very rapidly and very frequently! This is *not* what we want. We need the relay to remain in the transmit condition at normal Morse keying speeds, and then return to receive when the sending is complete. The combination of C2 and R3 achieves this.

The voltage appearing across C2 is fed to a VMOS field-effect transistor (FET), TR1, which acts as an electronic switch. When a voltage appears across C2, the FET switches on and passes a current through the relay coil, changing the contacts from receive to transmit. The diode, D3, across the relay winding protects the FET from being damaged by the large reverse-voltage spike which occurs across the coil when the FET switches off.

Construction

This is very simple – a circuit board is not required. The prototype was built into an old 50 g tobacco tin. All ground leads are soldered directly to the tin, thus supporting the components automatically. Phono sockets were used for the aerial, receiver and transmitter; you could use whatever connectors matched the rest of your station. The 12 V supply is fed through the tin using a 1000 pF feed-through capacitor, C3.

The 'dummy load', R1, must be able to withstand the RF output power of the transmitter. For novice use, 4 watts is adequate. Two suggestions for making up R2 from standard resistors are shown in **Figure 2**. Six 330 ohm, 1 watt resistors in parallel have a combined resistance of 330/6 = 55 Ω at 6 W. Two 100 ohm, 2 watt resistors in parallel have a combined resistance of 100/2 = 50 Ω at 4 W.

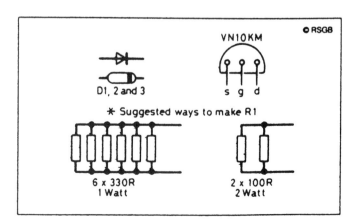

Figure 2 Component details

Make sure all the contacts to the relay are correct; also ensure that TR1, C2, D1, D2 and D3 are correctly wired.

Winding the toroid, L1, is very simple. It is made from PVC-insulated hook-up wire. Each time the wire passes through the centre of the toroid counts as one turn. The wire from the transmit socket to the relay simply passes through the centre! Use thin 50 Ω coax for the leads to the three sockets.

What happens to the receiver?

On transmit, the receiver is not connected to the aerial, but is only the separation of the switch contacts (about 0.5 mm!) away from the transmit lead. The receiver *will* pick up the transmitted signal, and there is a possibility that this signal will be enough to damage the receiver's sensitive input circuits. This can be prevented with the simple addition shown in **Figure 3**. Two diodes are connected back to back across the receiver socket. These act as a *limiter*, reducing the amount of signal that can enter the receiver. Solder these directly between the receiver socket and the tin.

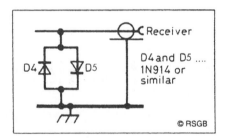

Figure 3 Receiver input protection

Warning

The wiring **must** be thoroughly checked. If the relay switch connections are wrong, there is a **great probability** that the full RF power will be applied to your receiver! You can check these connections as follows. With the 12 V supply connected, check that the relay energises (clicking sound heard) when the FET drain (d) is shorted to earth. Check the aerial and receiver connections with an ohmmeter or continuity tester. Disconnect the 12 V supply, and check that there is continuity between the centre pins of the aerial and receiver sockets. Reconnect the 12 V supply, short the FET drain to earth again (this will not damage the FET) and check that there is continuity between the centre pins of the aerial and transmitter sockets. If these tests show correct operation, disconnect the drain shorting wire (*most* important!) and your aerial auto-changeover is ready for use! Avoid using the device without the lid fitted.

Parts list

Resistors: 0.25 watt, 5% tolerance, unless otherwise stated
R1   50 ohms (Ω) – see text and Figure 2
R2   1 kilohm (kΩ)
R3   10 kilohms (kΩ)

Capacitors
C1   100 nanofarads (nF), 0.1 microfarad (μF)
C2   47 microfarads (μF) 15 V electrolytic
C3   1000 picofarads (pF) feedthrough

Semiconductors
D1, D2, D4, D5   1N914
D3   1N4001
TR1   VN10KM

Additional items
L1   T37–43 toroid
        Relay 12 V DPDT (DPCO) relay
        Three sockets to suit (phono, SO239, etc.)
        Thin coax cable and hook-up wire

42 A low-light indicator

Introduction

This is a simple one-evening project that can be built for the pure fun of it, or to use as the basis of a more complex project to switch your shack lights on when it gets dark! In its prototype form, it simply flashes an LED when the ambient light level drops to a preset point.

Operation

The heart of the circuit shown in **Figure 1** is a *photo-conductive cell*, also called a *light-dependent resistor* (LDR), a device whose resistance changes according to the amount of light falling on it. In bright light, the resistance is low (about 1 kΩ), whereas in the dark, its resistance is very high (up to 10 MΩ). The cell is made from a semiconducting material known as cadmium sulphide (CdS), and is enclosed in a small plastic container. The semiconductor is laid on a flat insulating surface in the form of a small flat ribbon. The ribbon construction gives a good area of surface for a given

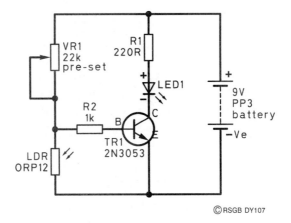

Figure 1 Circuit diagram

©RSGB DY107

length of ribbon, and the length of the ribbon is maximised by laying it out in a zig-zag pattern, as can be seen diagrammatically in **Figure 2**. In the dark, CdS is an insulator; when light falls on it, electrons are released inside the CdS, making it conduct. The more light there is, the more electrons there are, and the resistance falls.

In this circuit, the LDR is connected across the 9 V supply in series with a variable resistor, VR1. In this arrangement, the voltage that exists across the LDR will be determined by the light level. As the light intensity increases, the resistance of the LDR falls, dropping a smaller voltage across it. The

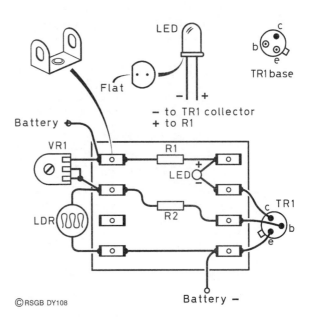

Figure 2 Component layout on tagstrip

©RSGB DY108

reverse happens when the light intensity falls – the voltage across the LDR increases. This voltage is used to drive an npn transistor, TR1, connected as an electronic switch. As the voltage on the base (B) rises, it will reach a point where TR1 will suddenly start to pass current, just as an ordinary switch does when pressed. The current flowing through TR1 also flows through a flashing light-emitting diode (LED), D1 and its series resistor, R1. D1 can be a steadily glowing type, if preferred. If you want to make the circuit switch the LED on at a different ambient light level, adjust VR1.

Construction

The prototype was made on an eight-tag tagboard (Figure 2). The resistors R1 and R2 can be laid on the tagboard for soldering, the rest of the components lying above or to the side of the board. Check the circuit after you have soldered everything on, then connect the battery. The LED should flash if you put your hand over the LDR, and VR1 can be adjusted to vary the point at which the LED lights. The project can be housed in a small plastic box with holes provided for the LDR and LED and an on/off switch if you want one.

Parts list

Resistors: all 0.25 ohm, 5% tolerance
    R1   220 ohms (Ω)
    R2   1 kilohm (kΩ)
    VR1   22 kilohms (kΩ) linear pre-set

Semiconductors
    LDR   ORP12
    TR1   2N3053 npn
    LED1   Flashing LED

Additional items
  Plastic box
  Tagboard with 8 tags
  PP3 battery connector

43 A J-pole aerial for 50 MHz

Introduction

For FM communication (i.e. voice and data) on the VHF bands, a vertical aerial is used to give all-round (non-directional) coverage. This is a half-wave aerial which can be fed at the end, thus removing the principal problem with the conventional vertical centre-fed half-wave dipole, which is that the feeder should leave the dipole at right angles. This is no problem when the dipole is horizontal, but can be difficult for the vertical dipole.

Basic facts

A feeder must be connected to an aerial at a point where the *impedance* (AC 'resistance') of the aerial closely matches that of the feeder. The difference between the two impedances gives rise to the voltage standing-wave ratio (VSWR), which is unity only when the two impedances are the same. With 50 Ω feeders, the feed point of a half-wave aerial is at the centre, where the aerial impedance is around the same value. At the end of a half-wave aerial, the impedance is high, so it is *not* a suitable point to connect a 50 Ω feeder.

Connection at this point can be effected using an RF transformer. RF transformers act in the same way as ordinary transformers, except that they are much smaller, and usually comprise wires of particular lengths adjacent to each other. **Figure 1** is a good starting point. It shows the aerial in its diagrammatic form. Notice that the aerial is in the form of an elongated letter 'J'; this shape gives rise to its nickname – the *J-pole*. The quarter-wave RF transformer is the lower 'U' section below the half-wave element. At the bottom of the U section, the impedance is zero (this may become clearer later) and at the top of the U section it is high, thus matching the aerial impedance. The coaxial feeder cable is connected part-way up the U section, where the impedance is around 50 Ω.

The practicalities

Figure 2 shows the aerial as constructed. As it is about 4.5 metres high, it may be too high for the average house loft, but is ideal for mounting outside, supported by a non-metallic pole or hung from a tree branch. The upper half-wave section is made from 1.5 mm insulated copper wire, as used in domestic mains wiring. The quarter-wave transformer below the half-wave section is made from 300 Ω balanced line ('ribbon cable'). The

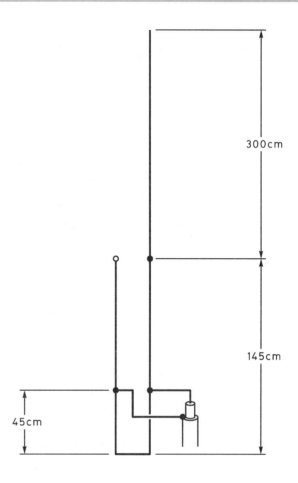

Figure 1 Overall dimensions of the 6 metre J-antenna

wires at the bottom of the transformer section are stripped of their insulation, twisted and soldered. At the upper end, only one wire of the balanced line is soldered to the bottom of the half-wave section. The other wire of the pair is not connected and is left insulated.

At the feed point of the transformer, the insulation needs to be carefully stripped from the balanced line. You will need a standing-wave meter (VSWR meter) in the coaxial line between your transmitter and the aerial, and you will need to adjust the position of the feed point. 45 cm from the bottom was the best point on the prototype, but this position is dependent upon the immediate surroundings of the aerial, and *must* be done when the aerial is in its final operating position. **Warning:** *Never* **make adjustments to the feed point when the transmitter is on. Make a VSWR measurement, switch off, move the feed point, switch on again, make another measurement, and so on.** You will need to aim for the lowest VSWR you can – certainly better than 2:1. Having found the best position, wrap all the exposed wires with self-amalgamating tape, to seal them against the ingress of moisture.

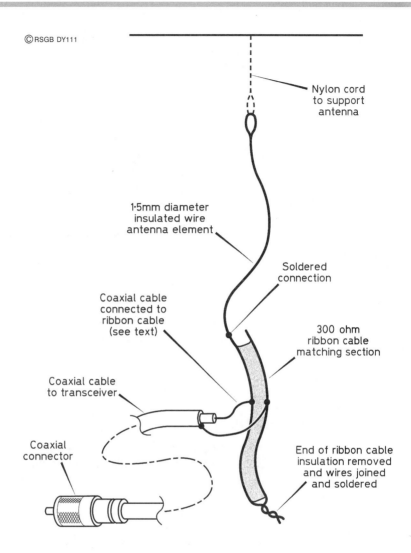

©RSGB DY111

Nylon cord
to support
antenna

1·5mm diameter
insulated wire
antenna element

Soldered
connection

Coaxial cable
connected to
ribbon cable
(see text)

300 ohm
ribbon cable
matching section

Coaxial cable
to transceiver

Coaxial
connector

End of ribbon cable
insulation removed
and wires joined
and soldered

Figure 2 Construction
details

How it performs

Figure 3 shows a computer prediction of how the J-pole radiates. It is
called a *polar diagram*, and shows the distribution of your transmitted
power when viewed 'from the end of your garden'. Most of your signal is
sent at a fairly small angle to the horizontal; very little signal goes upwards,
which is a good thing, of course. This also shows why the J-pole (or any
other vertical aerial) should not be called 'omnidirectional', which means
it radiates in all directions. It is omnidirectional *only* in the horizontal
plane.

151

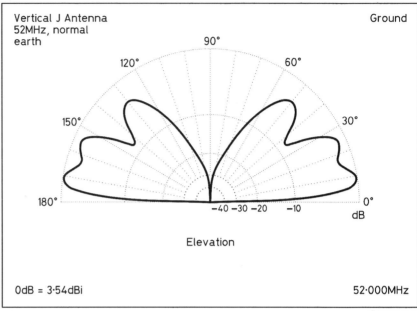

Figure 3 This polar diagram computed for the J-pole

Safety

Where you mount your aerial is a matter of personal preference and the restrictions of height and space, but the following safety rules *must* be applied.

1. Never fix an aerial where it may come into contact with power lines or telephone lines.
2. When climbing a ladder to put up an aerial outside, make sure the ladder is safe and that it is secured.
3. Don't do this alone. Preferably have someone with you. If this is not possible, make sure someone knows where you are.

Parts list

3.00 metres	1.5 mm insulated copper wire
1.50 metres	300 ohm balanced line
As required	50 ohm coaxial cable
As required	Self-amalgamating tape

44 Measuring light intensity – the photometer

Introduction

Before the days of automatic 'point-and-shoot' cameras, a photographer would use a light meter or *photometer* to measure the light level, then manually convert this reading into shutter speed and lens aperture settings to ensure a correctly exposed negative. Modern cameras have quite sophisticated photometers, which control the shutter speed and iris settings automatically.

A short explanation

A simple photometer circuit is shown in **Figure 1,** and is based on a device called a *light-dependent resistor,* or LDR. As its name tells us, its resistance depends upon the amount of light falling on it. In bright light, the resistance is relatively low (about $1\,k\Omega$), whereas in the dark, its resistance is very high (up to $10\,M\Omega$). The cell is made from a semiconducting substance known as cadmium sulphide (CdS), and is enclosed in a small plastic container. The semiconductor is laid on a flat insulating surface in the form of a small flat ribbon. The ribbon construction gives a good area of surface for a given length of ribbon, and the length of the ribbon is maximised by laying it out in a zig-zag pattern. In the dark, CdS is an insulator; when light falls on it, electrons are released inside the CdS, making it conduct. The more light there is, the more electrons there are, and the resistance falls.

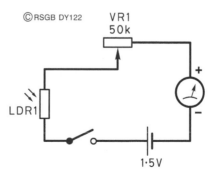

Figure 1 Circuit diagram of photometer

The photometer circuit

The circuit is simply a series connection of four things – the battery, the LDR, a variable resistor and a meter. A switch is also provided. The combination of the LDR resistance and that of VR1 determines the current flowing, which is indicated on the meter. Altering the resistance of VR1 sets the *sensitivity* of the photometer – you may want full-scale deflection of the meter needle for a bright light, or for a dim light.

It is simple to make, and a plug-in type matrix board is ideal to test the circuit, so that you can decide if you want to make a permanent version. Connect all the components in series; the only change you may want to make is the connection to VR1. If you find that the sensitivity control seems to work 'backwards', simply unsolder the wire from the end tag of VR1 and solder it to the opposite end tag. Problem solved!

In use

As soon as you connect up the battery, you will probably have a meter reading because of the daylight falling on the LDR. Shading it with your hand should reduce the reading. If the meter needle is hard over against the end-stop, turn VR1 until it indicates about half-scale. The LDR is very sensitive, and will read zero only in a dark room. If you put on a torch to see what the meter reading is, the LDR will detect the torch light, and will give a reading!

Here is a simple project where you can use the photometer in an experiment which has an analogy in radio. Draw a circle on a large (A3 or bigger) sheet of paper and divide it up into 30-degree sectors, as shown in **Figure 2**. Draw a series of smaller circles which divide the maximum radius into five. Look at the figure if you're not sure about this. Bring the LDR away from the circuit by using two long, flexible wires. The experiment must be performed in a darkened room (preferably in total darkness). Prepare a table with two columns, the left-hand one headed 'Angle (degrees)' and the right-hand one 'Meter reading'. Fill in the left-hand column 0, 30, 60 . . . and so on up to 360°.

Place the torch in the position shown, with its lens at the centre of the circle and pointing along the 0° line; switch it on. Place the LDR facing the torch and adjust VR1 until you have full-scale deflection of the meter needle. Suppose the meter indicates 10 units at this point. Enter this into your table in the 0° row. Keeping the torch the same distance from the circle centre, and pointing at it, move the LDR round all 30° positions and record the meter readings. Switch off the torch and take the sheet of paper and your tabulated results into daylight!

Lay the large sheet of paper on a table with your results beside it; then, at each 30° interval, plot the point along the radius corresponding to the meter

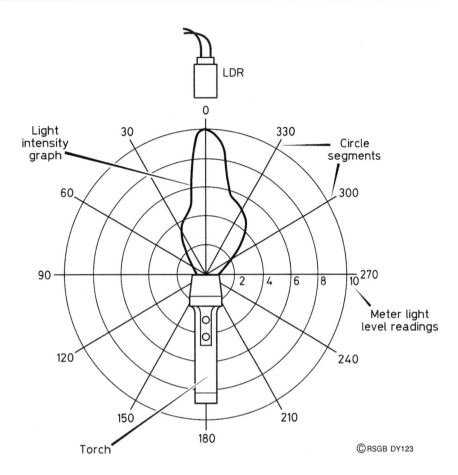

Figure 2 Torch light intensity pattern plotted at RSGB HQ

©RSGB DY123

reading. Then, join up all the points and you have what is called a *polar diagram* of light intensity. The use of the word 'polar' implies that the readings have been taken in a circle and plotted that way.

Light waves and radio waves are both examples of *electromagnetic radiation*. The torch is designed to 'beam' its radiation in a particular direction, just like an aerial does. Hence the use of the word 'beam' for a directional aerial. If a similar polar diagram is drawn for a Yagi-type aerial, it will show the same general characteristics as does Figure 2, namely a main direction (or 'lobe') where most of the energy is concentrated, with evidence of *sidelobes*, indicated by 'lumps' on the otherwise smooth main lobe.

Also in this book you will find a project for the construction of a UHF field-strength meter, which you could use to carry out a measurement of the polar diagram of a UHF aerial.

Parts list

Resistor
VR1   50 kilohms (kΩ) linear

Semiconductor
LDR1   ORP12

Additional items
Meter, 50 or 100 μA
Battery clip for single AA cell
AA cell

45 A 70 cm Quad loop aerial

Introduction

This is a description of how to make an aerial for the 70 cm band which has gain compared with the 'rubber duck' or the dipole aerial. It can easily be dismantled and reassembled, making it ideal for contest use.

The principles

Most aerials which comprise several similar *elements* arranged along a *boom* are variations of the design originated by Yagi and Uda, and which takes the name (for historical reasons) of the former, and is know as the Yagi array. Let us suppose we have a Yagi aerial beaming left to right in front of us. The elements get progressively shorter from the left (look at almost any TV aerial to see this). All the elements on a Yagi aerial are classified as follows:

- The *reflector* – the leftmost element as we look at the array. It is the longest. Next to it is:
- The *driven element* – this is the element connected to the feeder, which in turn runs down to the transceiver. It is slightly shorter than the reflector.
- All the elements beyond the driven element are called *parasitic elements*, or *directors*. They are shorter than the driven element and usually get progressively shorter as we progress to the right along the boom. The directors are mainly responsible for the *directivity* (or *beamwidth*) of the array.

The progression from a simple dipole (a driven element in isolation) to a Yagi array is simple, but is nevertheless important. To make an aerial of two elements, a reflector (*not* a director) is added to the driven element. For three or more elements, directors are added to the two-element design. Adding more and more directors soon becomes impractical, the reduction in beamwidth (such as it is) does not warrant the extra expense, weight and wind resistance that is incurred.

Instead of using linear (straight) elements, as in the generic Yagi, this design uses loops. Designs using squares of wire instead of loops are known as *Quad* aerials, and HF designs require large X-shape frames to support the large squares of wire. At 70 cm, however, the use of wire loops is easier, and they are self-supporting.

Construction

This is quite simple. Any type of material (metal, plastic, wood) can be used for the *boom* (the support for the elements) and for the mast. The elements are made from 14 SWG enamelled copper wire. 16 SWG hard-drawn aerial wire, which is not enamelled, has also been used with success. Thinner wire might result in a rather 'floppy' aerial! The separate parts of the aerial are held together with *jubilee clips* (hose clips).

The driven element is secured to the boom with a jubilee clip and a three-connector plastic connector block as shown in detail in **Figure 1**, and in the photograph. Cut the wire for the driven element 70 mm longer than the 700 mm indicated in Figure 1. Then, using sandpaper, remove the enamel from one end to a distance of 20 mm, and from the other to a distance of 50 mm. After forming the loop of the driven element, bend both stripped ends through 90°, and insert them into the first to holes of the plastic connector block (Figure 1). Do not tighten the screws yet. Push both ends into the block as far as they will go, then bend the 50 mm end back on itself and pull the ends back through the connector block so that the end you have just bent goes into the third hole in the block. Now tighten the screws in the block and in the jubilee clip.

Each director and the reflector should be made 40 mm longer than the circumferences shown in Figure 1. Strip the enamel, as before, from the last 20 mm at each end. Form the wire into the loop, slip the ends under the jubilee clip (Figure 1) and tighten it. You may find that it helps to solder the stripped ends together before securing the jubilee clip.

The boom is fixed to the mast using jubilee clips and wire, as shown in Figure 1. Solder the feeder cable to the driven element, with the braid soldered to the end of the driven element which is connected to the boom (this applies to metal booms; with plastic or wooden designs, the feeder connections are not critical).

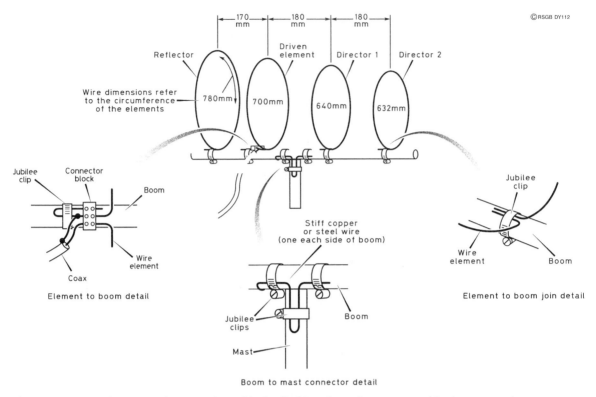

Figure 1 70 cm 4 element quad construction, with detail of how hose clamps are used in the construction

You will need a plug on the shack end of your feeder to suit the transceiver, aerial tuning unit (ATU) or the standing-wave meter (SWM) you are using. Always connect these in the order: transceiver – SWM – ATU – aerial.

Testing

Always test aerials outside and away from buildings (if possible!). This avoids getting misleading results.

Use a rubber duck, or whatever aerial you usually use, and tune around to find a repeater or beacon signal which is consistent. Note the reading on the S-meter. Then, connect your new Quad loop. Rotate it to give the strongest S-meter reading (don't forget it is directional). Verify that the meter reading varies as you rotate it. How does the S-meter reading compare with the original reading? Much depends on the siting of your original aerial; if you are comparing your Quad loop at ground level with a vertical on the chimney, you wouldn't expect your new aerial, even with its gain, to outperform a vertical which is well elevated!

Now a test on transmit is called for. The use of an SWR meter is essential here. Find a clear frequency and check that it *really is clear* before announcing your presence and carrying out the test. A reading of 1:1 is excellent, but any value less than about 1.8:1 would be acceptable. You can measure the directivity of your aerial using the field strength meter, also described in this series.

Materials

4 metres of 14 SWG enamelled copper wire
Material for boom and mast
15 amp connector block
7 Jubilee clips

The enamelled copper wire is available from AA&A Ltd, Sycamore House, Northwood, Wem, Shropshire SY4 5NN. Everything else is available from most hardware stores.

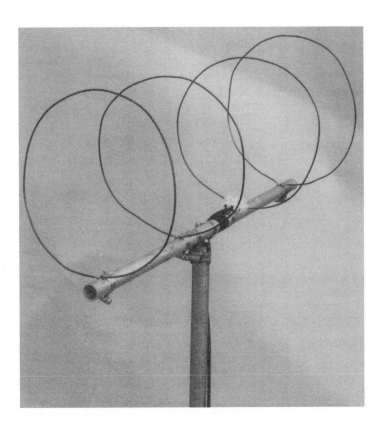

46 A UHF field strength meter

Introduction

It is always interesting, and often useful, to know just *where* the radiation is going from your aerial. How much of your radiated energy is going in the general direction of the station you are in contact with, and how much is being effectively wasted? Some of these questions can be answered with the use of a *field strength meter*. A field strength meter is simply a receiver, stripped down to its bare essentials, such that it responds only to the magnitude of the carrier. The use of a field strength meter assumes that the aerial under test is radiating a continuous carrier. Don't forget to find a clear frequency and identify your transmissions at least every quarter-hour, in order to comply with the terms of your licence.

Description

Two types of field strength meter are shown in **Figure 1**. You will recognise both circuits (Figure 1b particularly) as being types of 'crystal set' with a meter replacing the headphones. Figure 1a is a broad-band HF design (there is no tuning provision) and Figure 1b is tuned in the same way as the crystal set; with a loop of wire as an aerial, it will perform well in the VHF/UHF range.

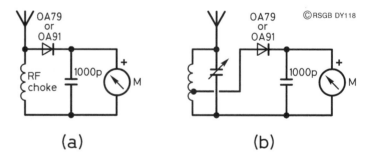

Figure 1 Construction of the UHF field strength meter

A tuned field strength meter also doubles as an *absorption wavemeter* if it is well calibrated. Such devices are useful for detecting transmitter harmonics also. This design is intended for use in the UHF band, so it will have to be sensitive around 432 MHz.

Construction

The field strength meter comprises a loop of wire, 600 mm long, which acts both as aerial and as the tuning inductance, a diode, a capacitor, a connector block, a meter and a length of twin wire. All the components, with the exception of the meter, are fixed to a pole with a jubilee clip, as shown in **Figure 2**. The meter should have a sensitivity of between 50 μA and 100 μA, or a multimeter can be used. The multimeter is more flexible, as you can select different current ranges, giving you a range of sensitivities.

Using the field strength meter

Connect your meter to the ends of the twin wire from the pole. Place a hand-held transceiver about 2 metres away, and press Transmit. If there is no meter reading switch off the transmitter and check the wiring. If the needle attempts to go negative, simply reverse the wires to the meter. If the reading is too high, either move the transmitter further away, or increase the current range on the multimeter. Try changing the orientation of the transmitter aerial, and note how the signal varies.

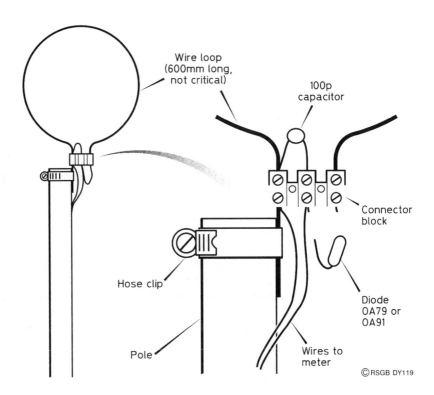

Figure 2 Assembly of the field strength meter

©RSGB DY119

161

To measure the *polar diagram* (a graph of the signal strength against the angle between the aerial boom and the field strength meter, plotted on polar axes) of a beam aerial, mount the loop as far away from the aerial as you can, preferably at the same height, and rotate the aerial, say 15 degrees at a time, and record the signal strength, until the aerial has been turned through 360 degrees. If you need help plotting the graph, enlist the help of a friend who has done it before, or one who knows about polar graphs! If you have already built the Photometer project, you will have measured and plotted the polar diagram of the light intensity from a torch. Now you will see the great similarity!

Parts list

Capacitor	100 picofarads (pF)
Diode	Germanium, OA79 or OA91
Connector block	10 A, 3-way
Wire	600 mm of 16 SWG enamelled copper
	Length of twin cable for meter connection
Clip	Jubilee (hose) clip

47 Christmas tree LEDs

Introduction

A novelty ideal for the festive season, this circuit causes one LED at a time to light up around a small cardboard Christmas tree.

Warning

This circuit uses members of the integrated circuit family known as CMOS (complementary metal-oxide semiconductor). These use very little current and can be completed destroyed if they come into contact with the magnitudes of static electricity that most of us carry about when we walk on carpets and wear rubber shoes. You will never know if this wanton destruction has happened – all you will discover is that your circuit doesn't work and that you have tested *everything*. To avoid this problem do the following things:

1. Before you open the little packet in which each IC is supplied, touch something which you *know* to be earthed – the metalwork of any equipment which is mains-earthed, for example. Then open the packet.
2. Let the IC fall gently on the bench – don't pick it out with your fingers. Touch your earthed metalwork again. Pick up the IC and insert it gently into its holder.
3. Repeat the process with the second IC.

The circuit is safe from destruction while it is connected to the battery. However, when the battery is removed, the same care should be exercised with its handling, because there are no supply decoupling capacitors across each IC.

Description

Figure 1 shows the layout, and how the wires are connected from the circuit board to the LEDs around the tree. The circuit, shown in **Figure 2**, is quite complicated, so you need to be confident in your logical approach to circuit-building before you attempt this one! It uses two common integrated circuits (ICs). IC1 is simply an oscillator which provides timing pulses for IC2, which 'counts' up to a maximum of 10. The outputs of the counter are indicated by light-emitting diodes (LEDs); red, orange, yellow and green are common colours which you can use.

The circuit is built on a single piece of Veroboard measuring 29 holes by 12 strips. **Be aware that there is no strip labelled 'I', so don't make mistakes in your counting!**

First of all, cut the tracks in the positions shown in Figure 1 using a 3 mm ($\frac{1}{8}$ inch) twist drill rotated between thumb and forefinger. Then, solder in the IC sockets, the notches facing row M. Wire up and solder in the links and then the Veropins for the connections to the LEDs and battery. Having done this, the resistors and the capacitor should be fitted. When wiring the LEDs, each cathode (the lead adjacent to the 'flat' on the LED encapsulation) is connected to R4, and the anodes go to separate pins on IC2.

Testing

Hold up your circuit board to the light and check carefully for solder bridges between adjacent tracks. Then check again that the wiring is correct. Place the ICs in their holders, with the notched ends lining up with those on the holders. Connect the battery, and the LEDs should illuminate in sequence.

If you have no success, you are now wishing you had checked the circuit more carefully! Learn something from your mistake and it will not have

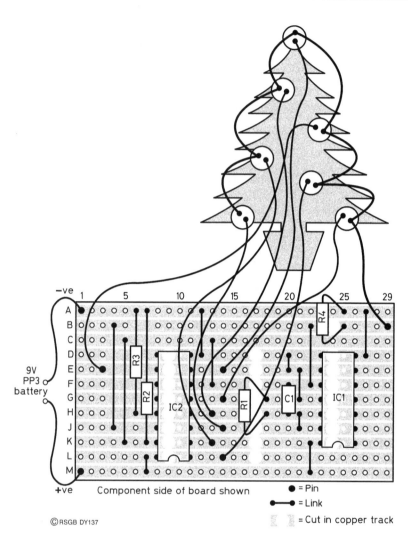

Figure 1 Christmas tree, component layout

©RSGB DY137

• = Pin

•—• = Link

〰 〰 = Cut in copper track

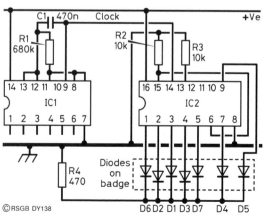

Figure 2 Christmas tree, circuit diagram

©RSGB DY138

been in vain! The first thing to check is that the LEDs are the correct way round. If that is OK, then check the positions where the tracks are broken (intentionally!). After that, are the wire links in the right places – had you forgotten there is no row 'I'? Check *all* the wiring, then check *again* for solder bridges, and switch on again. One of these tests should have revealed a fault. If it *still* refuses to work, perhaps one (or both) of your ICs were damaged by static electricity, despite your precautions – or did you choose to ignore them?

Parts list

Resistors: all 0.25 watt, 5% tolerance
R1	680 kilohms (kΩ)
R2, R3	10 kilohms (kΩ)
R4	470 ohms (Ω)

Capacitor
C1	0.47 microfarad (μF) min. metallised polyester film

Semiconductors
IC1	4011
IC2	4017

Additional Items
D1–D7	5 mm LEDs in choice of colours
	14-pin DIL socket for IC1
	16-pin DIL socket for IC2
	PP3 battery clip and battery
	Veroboard, Veropins
	Insulated wire for links

48 An audio signal injector

Introduction

An audio signal injector is a device used to test audio frequency circuits. It is simply an oscillator running at a frequency in the audio range, so that when its output is fed to the input of an amplifier, it will produce a sound in the loudspeaker if the amplifier is working. The oscillation is so rich in harmonics that the signal can also be heard (sounding rather different) when injected into an RF circuit.

The design

The circuit is shown in **Figure 1**, and is a basic *astable multivibrator*, a free-running oscillator producing a roughly rectangular-wave output. The two transistors, TR1 and TR2, operating as switches, switch on and off alternately at a frequency around 500 Hz. The prototype was constructed on plain matrix board (no copper strips), as illustrated in **Figure 2**.

Both transistors are type BC108, which are only a few pence each new, and can be found at almost any rally. You can add an on/off switch, or simply disconnect the battery when you are finished using it. To make the unit in one piece, the battery can be taped to the board, as the diagram shows.

The probe itself is made from a short piece of stiff insulated wire, soldered to a tag on the board; an earth lead is also soldered to the board, and terminated in a crocodile clip to attach to the ground lead of the equipment under test.

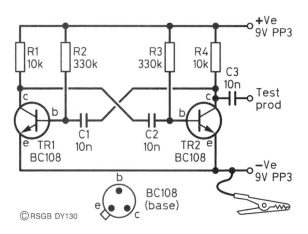

Figure 1 Circuit diagram of the signal injector

© RSGB DY130

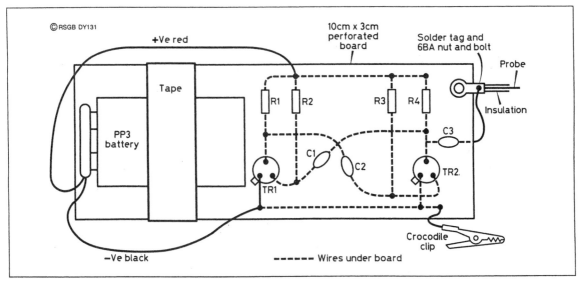

Figure 2 Component layout of the signal injector

Using it

After checking the wiring of the circuit, connect the battery. Find a transistor radio that is known to work. Carefully remove the case, switch on and apply the probe to the centre tag of the volume control. A *very* loud buzz should be heard from the speaker, indicating that the audio circuits of the radio are working. Using the injector to fault-find equipment you have made yourself is rather more instructive and rewarding, because you *know* where to inject the signal, and you should *know* what to expect when you do.

Warning

Do *not* work on any equipment connected to the AC mains. Work only on battery-powered circuits, for your own safety.

Parts list

Resistors: all 0.25 watt, 5% tolerance
    R1, R4   10 kilohms (kΩ)
    R2, R3   330 kilohms (kΩ)

Capacitors
C1, C2, C3   10 nanofarads (nF) or 0.01 microfarad (μF) ceramic

Semiconductors
TR1, TR2   BC108

Additional items
  PP3 battery and connector
  Matrix board 10 cm by 3 cm
  6BA solder tags
  Thick insulated wire for probe
  Crocodile clip

49 Standing waves

Introduction

Everyone has, or *should* have, a standing-wave ratio (SWR) meter as part of his/her array of test gear. Most people know how to use it, but what does it really do?

Before attempting to answer that question, we need to look at some aerial fundamentals. An aerial is a *transducer*, the word meaning 'to lead across'. We use it whenever one form of energy is converted into another form. A bulb is a transducer; it converts electrical energy in the filament to radiated heat and light energy. **Figure 1** shows the situation.

An aerial, or antenna, is also a transducer; it converts radio-frequency (RF) energy in the feeder (or *transmission line*, to give it its proper name) into radiated electromagnetic energy, in the manner shown in **Figure 2**. The aerial has resistance, just like the bulb filament, and if the filament had zero resistance, there would be no radiated energy. In the same way, if the aerial had no resistance, there would be no radiation from it.

Transmission lines should convey RF energy from transmitter to aerial with the minimum power loss. A common form of transmission line is the *coaxial*

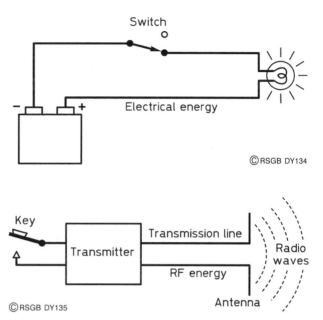

Figure 1 The light bulb converts electrical energy to electromagnetic light energy

Figure 2 The antenna converts RF current to electromagnetic radio waves

cable. As you might expect, any feeder cable has a DC resistance caused by the resistance of the copper wire from which it is made. It also has an AC resistance, caused by the capacitance between the centre conductor and the braid, and by the inductance of the cable itself. This means that the feeder has an *impedance*, which is constant for the particular type of cable. This is what we call the *characteristic impedance* and, for the cables used in most amateur radio applications, it is 50 Ω.

If this impedance can be made the same as that of the aerial (it is already the same as the impedance of the transmitter output), then the transfer of energy will occur with minimum loss. If the aerial and cable impedances are *not* the same, then there is a *mismatch*, which causes some of the RF energy to be reflected back towards the transmitter. We now have a situation where RF energy is flowing along the cable from the transmitter to the aerial (the *forward* wave) and, at the same time, flowing from the aerial to the transmitter (the *reflected* wave). The two waves interact along the cable and form a stationary pattern of voltage and current. The pattern is known as a *standing wave*, and can be visualised from the waveforms in **Figure 3**. The ratio of the maximum voltage to the minimum voltage on a given wave defines the *voltage standing-wave ratio* (VSWR), or just *standing-wave ratio* (SWR) for short.

The SWR meter is easy to use. It is positioned in the feeder between the transmitter and the aerial tuning unit (ATU) if there is one, or between the transmitter at the aerial otherwise. Most meters have a single meter and a four-position switch. The transmitter is first keyed and, with the switch in

169

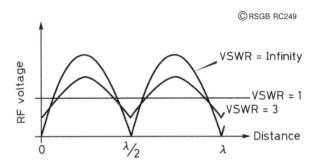

© RSGB RC249

Figure 3 Standing waves on a transmission line

the 'calibrate' position, the sensitivity control is adjusted to give full-scale deflection on the meter. The switch is then changed to 'forward' to read the forward power, to 'reflected' to read the reflected power, and to 'SWR' to read the value of the standing-wave ratio. When there is no reflection (see Figure 3), the meter should read 1:1 or, simply, 1.

Most SWR meters remain in the feeder line while the transmitter is operating, so the condition of the aerial and feeder can be constantly monitored. Problems with the aerial (such as water entering the feeder at its junction with the driven element) are immediately shown up. Without the use of the SWR meter, the situation would slowly deteriorate over several months and you would be left wondering why so few stations were answering your calls!

50 A standing-wave indicator for HF

Introduction

The standing-wave ratio (SWR) meter shows how well the aerial system, including the feeder, is *matched* to the output of the transmitter. This design does not *measure* SWR, but it gives an indication of when the SWR is minimum for a given system of aerial and feeder. The design is usable on the HF bands from 1.8 to 28 MHz, and can be used at 50 MHz with reduced sensitivity.

How it works

There are two types of wave in any feeder: the *forward* wave, which travels from the transmitter to the aerial; the *reflected* wave, which travels back to the transmitter from the aerial. The presence of a reflected wave is evidence that some of your transmitted power is not being radiated, but is being returned to the transmitter to be lost as excess heat. When aerial and feeder are perfectly matched, there is *no* reflected wave, and all the power from the transmitter is radiated.

Referring to the circuit of **Figure 1,** a tiny fraction of the signal is removed by the transformer, T1, and by the capacitors, VC1 and C1. It is then detected by the germanium diodes, D1 and D2, and any residual RF removed by the capacitors, C2 and C3. The currents through the diode and meter (depending on the position of switch, S1) represent the forward and reflected signals. VR1 acts as a sensitivity control for the meter.

It pays to shop around for a suitable meter. Surplus types from tape recorders and hi-fi equipment are usually ideal for this purpose. A new one would cost several pounds. The more sensitive the meter, the more sensitive your indicator will be. Meter sensitivity is given by the current required to give full-scale deflection (FSD) of the pointer. One with an FSD of between 50 and 200 micro-amps (μA) is suitable for this circuit. The higher the FSD, the less sensitive the circuit.

Construction

The meter circuit and the sampling transformer (see **Figure 2**) are built and mounted on Veroboard of the copper-strip variety. It simplifies construction

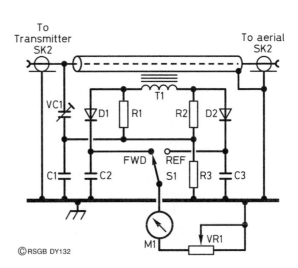

Figure 1 Circuit diagram of the SWR meter

©RSGB DY132

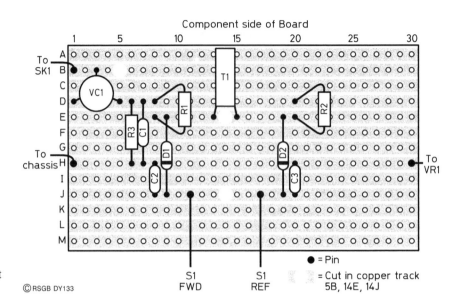

Figure 2 Component layout of the SWR meter

© RSGB DY133

● = Pin

▨ = Cut in copper track 5B, 14E, 14J

S1 FWD   S1 REF

but reduces the operational range of the meter to below 30 MHz because of the capacitive coupling between strips. The board has 13 strips by 30 holes, although you can reduce this if you have a smaller case.

Firstly, cut the tracks at the three points shown. Then insert and solder Veropins for connections to the external components, the switch, variable resistor and the meter. Solder in the components starting with the resistors and followed by the capacitors and the diodes, ensuring that the diodes are inserted correctly.

Now you have to wind the transformer, T1, on a small toroidal ferrite core. Wind the secondary with 15 turns of 36 SWG enamelled copper wire, spaced evenly over about two-thirds of the former. The turns should not overlap, and considerable care must be taken; the wire is very thin, will kink easily and will break if you apply too much tension. The 'primary' is an 8 cm length of 50 Ω coaxial cable which passes through the toroid on its way between the input and output connectors. The braid of the cable is connected to the case at only *one* of the connectors (see Figure 1); this prevents the screen and the metal case between the two sockets forming a single, shorted turn.

The ends of the secondary winding must be carefully stripped of their enamel with sandpaper, before attaching the toroid to the board with cotton or nylon fishing line. On no account must wire be used for this!

Solder the secondary connections of T1 to the board and thread through the coaxial cable ready for soldering to the connectors.

The case used was an aluminium box (Maplin LF02C), but any suitable metal box could be used. Aluminium is preferable, as it is easily drilled with simple tools. Use standoff insulators to mount the board in the case. Once this has been done, the leads from the board to the chassis-mounted components can be soldered. So can the coaxial cable passing through the toroid. Make the lead from the input socket to VC1 as short as possible.

Setting up

You will need a 50 Ω dummy load and a transmitter to set up your indicator. Connect the transmitter to SK1 and the dummy load to SK2. Set the toggle switch, S1, to *forward* and the sensitivity control, VR1, to mid-travel. Switch on the transmitter, and set VR1 for maximum meter deflection. Switch to *reflected* and adjust VC1 until the reading is minimum (ideally zero). This completes the setting up!

Using the indicator

For setting up an aerial, connect your circuit between the transmitter and the cable leading to the aerial. With S1 in the *forward* position, key the transmitter and adjust VR1 for maximum reading on the meter. Switch to *reflected*, and then adjust your ATU to give minimum reflected power. If your adjustments are to be made to the aerial itself, to give minimum reflected power, you must make a note of the reflected reading, switch off the transmitter, make a change to the aerial, key the transmitter, and note whether the reflected power is greater or less than before. Then, make more changes to the aerial. **Never adjust your aerial with the transmitter on.** Make your adjustments on an unused frequency, and do it as quickly as possible, thus avoiding (or minimising) interference to other stations.

Parts list

Resistors: all 0.25 watt, carbon 5% tolerance
(or Maplin 0.6 watt metal film)

R1, R2	27 ohms (Ω)
R3	2.2 kilohms (kΩ)
VR1	10 kilohms (kΩ) linear

Capacitors

C1	220 picofarad (pF) disc ceramic 50 VDC
C2, C3	0.1 microfarad (μF) disc ceramic 50 VDC
VC1	20 picofarad (pF) trimmer

Semiconductors
D1, D2   OA91 germanium

Additional items
S1   Single-pole changeover (SPDT or SPCO)
SK1, SK2   Coaxial sockets to suit station standards
                 Veroboard – 13 strips by 30 holes
                 Veropins (7 off)
                 Amidon FT 50–43 ferrite toroid
                 Meter, less than 200 µA FSD
                 36 SWG enamelled copper wire
                 Short length of UR43 or RG58 coaxial cable
                 Insulated stranded wire
                 Aluminium box
                 Standoff insulators for mounting the board
                 Knob for the sensitivity control

51 A moisture meter

Introduction

Dry rot (*Merulius Lacrymans*) can strike havoc in buildings, causing the timbers to decay and crumble to dust – hence the term *dry rot*. Wood is attacked only if its moisture content rises above 20%.

Construction

The circuit of the moisture meter is shown in **Figure 1**. The two probes touch the wood, and the current that flows between them depends on the moisture content of the wood. If the moisture is sufficiently high, the current, after amplification, will be enough to light the LED.

The meter can be made on a piece of plain matrix board (no copper strips), as **Figure 2** shows. The board is big enough (10 cm by 2.5 cm) to accommodate the PP3 battery, taped on. No case is needed, unless you want

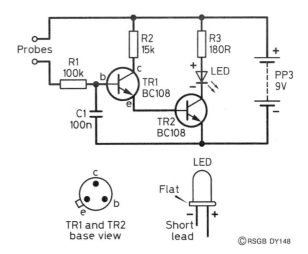

Figure 1 Circuit diagram of a moisture meter

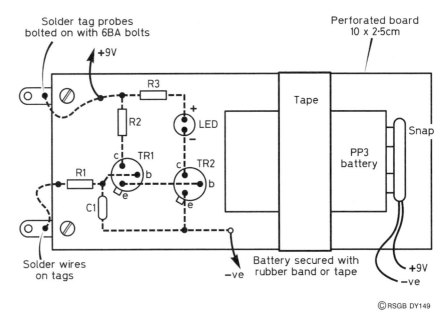

Figure 2 Moisture meter, component layout

to leave the meter in a damp location for a protracted period! Make sure that the transistors and LED are mounted correctly. In Figure 2, the connections as shown to TR1 and TR2 are illustrated as if the transistors were transparent. An on/off switch is not really necessary, as only a very small current flows when the probes do not touch anything. Use solder tags, screwed to the board, to act as probes.

After the assembly is completed, check your circuit one final time, and then connect the battery. Nothing should happen at first. If you lick your forefinger and hold it across the probes, the LED should light.

Using it

The prototype was compared with a commercial moisture meter, and the LED lit when the moisture was around 20%. This was quite fortuitous, as the point at which the LED lights depends both on the separation of the probes and on the gain of the two transistors.

In addition to searching for dry rot, the instrument may be used to monitor the moisture in the soil of household plants. In this case, probes made of 16 SWG copper (*not* enamelled) should be soldered on to the two tags, and should penetrate the soil to a depth of several centimetres, and R1 may require adjusting so that the LED extinguishes if the soil is too dry, and lights if the soil is sufficiently moist. If you wanted to leave the meter with the probes in the soil, an on/off switch *would* be necessary.

Parts list

Resistors: all 0.25 watt, 5% tolerance
R1	100 kilohms (kΩ)
R2	15 kilohms (kΩ)
R3	180 ohms (Ω)

Capacitor
C1	0.1 microfarad (μF) polyester

Semiconductors
TR1, TR2	BC108
LED	Any shape or colour will do

Additional items
PP3 battery and connector
Solder tags (2 off) for probes
Matrix board 10 cm by 2.5 cm

52 Simple aerials

Introduction

The performance of *any* receiver or transceiver, no matter how expensive it is, is limited by the aerial that feeds it. Two of the most frequently asked questions are:

- Which is the best sort of aerial to use?
- Where is the best place to locate an amateur radio aerial?

To answer these questions, you must ask yourself what sort of operation you want to do. Are you interested in local, chatty contacts on the lower bands or VHF, or are you more disposed towards long-distance (DX) contacts, and on what band?

A house with a moderately sized garden is assumed in the diagrams here, to illustrate the configurations of some simple aerials. You would not need all these aerials festooned around your house, because one or two would be sufficient for your needs. The problems incurred by properties with more restricted space will be covered later.

VHF aerials

For VHF operation, the aerial should be mounted as high as possible, either on a mast or on a chimney. For all-round coverage on FM and the local repeaters, a vertical colinear is a good choice. For SSB and CW DX operation, a horizontal rotatable beam is needed. If satellite working is envisaged, you will need to contemplate mounting an elevator on top of your rotator, so that your beam can point in *any* direction, including vertically upwards! An advantage of satellite working is that the aerials do not *necessarily* have to be up in the air, provided you have a relatively uncluttered site. Your rotator and elevator can be at ground level, which is good!

If the VHF aerial is mounted on the chimney, use a double mounting bracket, particularly if you have a beam and rotator. Keep the TV, broadcast FM and amateur aerials as far apart as possible, and keeping the feeders separated is also a good plan.

The dipole aerial

One of the simplest types of aerial for single-band operation is the half-wave dipole. (The name 'dipole' simply means 'two poles' or 'two elements', and

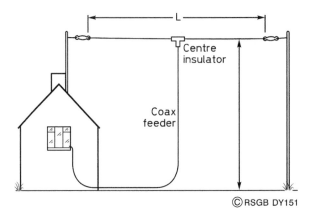

Figure 1 Layout for a dipole aerial

©RSGB DY151

in this case the total length of the dipole is approximately half a wavelength at the operating frequency.) It is usually fed in the centre by coaxial cable as shown in **Figure 1**. The length of the dipole for the lowest frequency in each band is shown in **Table 1**. Normally, the length of the aerial will be 'trimmed' to be tuned to the centre frequency of the part of the band in which you will operate. This is done using the data in the right-hand column of Table 1. As an example, suppose you wanted your aerial to be resonant at 3.7 MHz. The table gives an overall dipole length of 42.86 m for 3.5 MHz. To resonate the aerial 200 kHz *higher*, then this length must be shortened by 2 × 0.595 m = 1.190 m. Your dipole would thus be 41.67 m long. Remember to allow extra wire for fixing the dipole ends to the insulators.

Table 1 Dipole lengths for lowest frequency of each band and the length to be trimmed from each to raise the resonant frequency by 100 kHz

Band (MHz)	Dipole length (m)	Trim each end (mm/10 kHz)
1.8	83.33	2190
3.5	42.86	595
7	21.43	150
10	14.85	70
14	10.71	35
18	8.33	20
21	7.14	15
24	6.03	12
28	5.36	10
50	3.00	6

On the lower-frequency bands, the lengths become rather large. In this case, you can 'bend' your dipole, as illustrated in **Figure 2**. The length of wire required to give an acceptable value of SWR (less than 2:1 on transmit) may need to be different from the calculated value, so be prepared to experiment!

Dipoles are single-band aerials, although they will often work acceptably on the third harmonic of their design frequency: a 7 MHz dipole often operates reasonably well on 21 MHz. It is possible to operate several dipoles in parallel, as **Figure 3** shows. Interaction between the elements can occur if the spacing between them is less than about 10 cm. A multi-band dipole, as shown in Figure 3, has the elements separated with plastic spacers, and drooping ends to produce maximum spacing between the elements' ends.

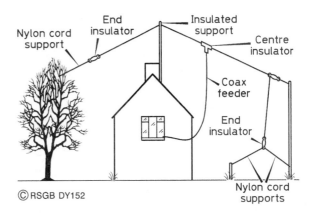

Figure 2 Possible layout for a dipole aerial in a confined space

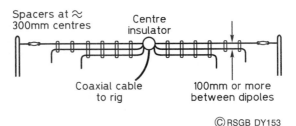

Figure 3 Multi-band dipole aerial

The long-wire aerial

This aerial is simple, cheap, easy to erect, and suits most houses and gardens, as **Figure 4** shows. Using an aerial tuning unit (ATU), an end-fed long wire can function on several bands when used with a set of radials or a counterpoise. **Figure 5** illustrates the setup. The length of the aerial will determine the bands which will be covered.

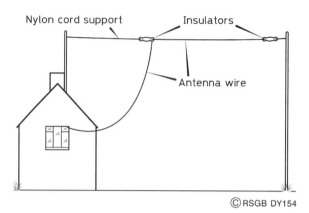

Figure 4 Long-wire or inverted-L aerial

©RSGB DY154

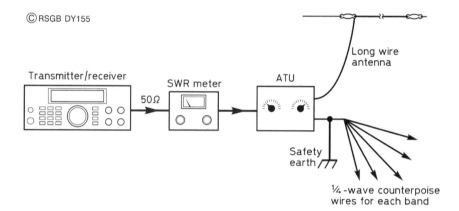

Figure 5 How to connect a radio to a long-wire aerial

A wire length of 10.5 m will work on the 40, 30, 17, 15 and 12 m bands.

A wire length of 15.5 m will work (with an ATU) on the 80, 40, 20 and 12 m bands and possibly (depending on your ATU) on the 17 and 15 m bands.

A wire length of 26.5 m will operate on all bands, but may be difficult to load on 10 m.

The wire lengths given here may need some adjustment because of the geometry of your particular house and garden. For receive-only purposes, the lengths are far less critical.

In general, you cannot get a good radio-frequency (RF) earth from a first-floor (or higher) shack. Unless a good RF earth exists within a small fraction of a wavelength of the transceiver, an *artificial ground* comprising a single $\frac{\lambda}{4}$ radial or counterpoise will be needed. You will need one counterpoise for each band you intend to use, and the wire can be concealed around the skirting-board of the shack, or under the carpet. Make sure that the free end

Table 2 Lengths of elements for vertical antennas, radials for verticals and counter-poises for end-fed long wire antennas

Band (MHz)	Element length (m)
1.8	39.66
3.5	20.40
7	10.20
10	7.14
14	5.1
18	3.96
21	3.4
24	2.95
28	2.55

of each counterpoise is well insulated; this point can carry a *very high voltage* when you transmit; anyone coming into contact with this can suffer very severe RF burns. Counterpoise lengths can be read from **Table 2**.

The vertical aerial

The single-band vertical aerial is sometimes used by DX operators because it has a low angle of radiation, which favours long-distance propagation. However, it must be sited clear of obstructions and must have a *good* counterpoise or radial system. Illustrations of the vertical aerial are shown, and the lengths of the vertical and radial sections are given in Table 2. The centre of the coaxial feeder is connected to the vertical section, and the braid to the counterpoise or radial system, which is made up of four or more wires buried just below the surface and joined together near the base of the aerial.

Cable entry to the house

Bringing coaxial cable into the house by an open window must be regarded as a temporary measure. Wooden window frames can be drilled, one hole for each feeder. Make the holes slope downwards from inside to out to prevent rain entering, and treat these with wood preservative. Leads from long-wire and inverted-L aerials should be kept separate from other cables.

Alternatively, a plastic pipe large enough to take all your feeders could be fitted into the brickwork (again, sloping downwards towards the outside). You may want to let a friendly builder do this for you.

53 A breadboard 80 m CW transmitter

Introduction

In the early days of radio, many circuits were built on a wooden baseboard, the parts being screwed down on the board. This was called breadboard construction, because it was a breadboard that was frequently commandeered for the process! Wives have always been generous in this respect, it appears!

This circuit was originally designed by GM3OXX, and became known as the *Oner*, because it was built on a circuit board one inch square! The circuit appeared in the G QRP Club journal *Sprat* and, since that time, many hundreds of *Oner* circuits have been built and used on the air. It is a well-proven circuit.

The transmitter has no tuned circuits in the power **amplifier** (PA) and thus has a rather high *harmonic* content. It **must** be used with the *low-pass filter* described elsewhere in this book. Without the low-pass filter, interference will be caused to other stations.

Simple aerial changeover switching is provided, which allows this circuit to be used with any of the 80 m receivers, such as the *Colt*, described in this book. It can also be used with any kit or commercial receiver for the 80 metre band.

The circuit

The transmitter circuit is shown in **Figure 1**. TR1 is a crystal oscillator, the frequency of which is controlled by crystal X1. A small trimmer capacitor, TC1, is added to allow the frequency of X1 to be varied by a small amount. If adjustment of this trimmer is made possible from the front panel, it is useful to adjust the transmit frequency to avoid other stations already on the crystal frequency. The collector load resistor, R2, of the oscillator transistor, TR1, determines the power output; a value of 3.3 kΩ seems to work well in producing an output of 3 watts.

TR1 is directly coupled to TR3, a VMOS transistor (a type of field-effect transistor (FET)). This acts as the power amplifier (PA) stage. TR3 should

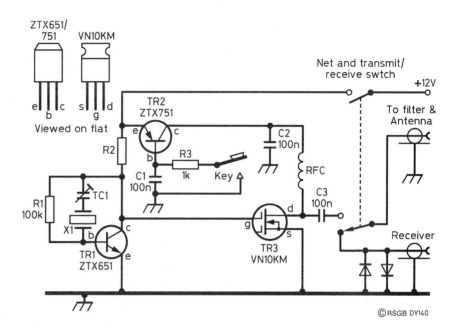

Figure 1 Circuit diagram of the breadboard transmitter

give about 3 W output, which is then coupled to the output by C3. The radio-frequency choke (RFC) providing the drain load of TR3 is simply a few turns of wire on a ferrite bead.

TR2 is an interesting addition to the circuit. It is used as a switch to 'key' the PA, TR3. The transmitter *could* be built without TR2, just placing the Morse key between the top of the RFC and the 12 volt supply. Adding TR2 is helpful, because it means that one side of the Morse key can be grounded (always a good thing), and some degree of *shaping* of the output RF waveform is provided by R3 and C1. This makes the transmission sound a little better and reduces the possibility of spurious frequencies being generated and transmitted. TR2 is a pnp transistor; note that it is the *emitter* of this transistor which is connected to the positive side of the supply.

Some form of changeover switching is needed for the aerial. A double-pole changeover toggle switch can be used. See the chapter on switches, later on in this book. One pole is used to switch the aerial between transmitter and receiver; the other pole is connected in the 12 volt supply line, and is labelled RECEIVE/TRANSMIT – NETTING. Its use will be described later. In this simple circuit, the PA cannot work when the key is open, because the key breaks its supply (via the RFC). When the key closes, TR2 switches on and applies the 12 volt supply to the top of the RFC. C2 is a *decoupling* capacitor, which prevents any residual RF signals at the top of the RFC reaching TR2.

Building

The prototype was put together as follows. Take a piece of plain printed-circuit board (PCB) measuring 5 cm by 4 cm. Then, with a *new, sharp blade* in a junior hacksaw, draw the blade horizontally across the surface of the copper in order to make a pattern of 6 squares along the 5 cm side and 5 squares along the 4 cm side. No more pressure should be applied than is necessary to cut through the copper! All the parts will be soldered on these pads in a form of surface-mount construction. To do this, each active pad (i.e. one that is going to have a component soldered to it) needs to be *tinned*. This means coating the pad's surface with solder, and is carried out as follows. Place the hot tip of the soldering iron on to the pad, and hold it there for a second or so. Then, with the tip still in place, touch the end of your reel of solder *on the pad*, not the tip of the iron. The solder should flow evenly all over the pad, and you can remove the iron. The solder should solidify in a rounded, shiny blob! This provides a good surface for making soldered joints.

To join component leads to the pads, cut each lead about 1 cm long, and then bend the last 2 mm at right angles to the rest of the lead. As you did before, tin the 2 mm length of each lead. Place the tinned portion on to the pad, and place the tip of your iron *on the pad*, close to the lead. The solder on the pad and on the lead will melt and run together; remove the iron and *hold the component still* until the solder solidifies. When the joint has cooled, give the lead a gentle tug to make sure you have a good joint (a good *mechanical* joint is usually a good *electrical* joint, too!). Each transistor straddles three pads, so the centre lead will need to be shorter than the other two. Take care here to get the lead lengths right – if you do, you will be surprised how much more firmly the transistor is held than if you just botched the lead lengths by bending them to fit! Make sure the connections to the transistors are correct.

Winding the RFC is quite simple. Seven (or more) turns of thin (32 SWG) enamelled copper wire are threaded through a small ferrite bead. This requires care, because the bead is small and the wire is thin. Trim the ends to within about 1 cm of the bead, remove the enamel carefully with sandpaper and tin the bare ends, prior to soldering the choke to the board.

After completion of the wiring, check the circuit against Figure 1. Breadboarding a circuit like this has its advantages, but it can have disadvantages, too. One of these disadvantages is that it can make circuit checking difficult. For a simple circuit like this, it is not too bad! Check that no solder has run between the pads. Plug in your crystal for the 80 m band, and connect up the 12 V power supply. Do not connect the Morse key yet and do not switch on the power.

Testing and operating

Clear your workbench of all metallic objects, slivers of copper, bits of wire, etc., switch on your power supply. With an external receiver, listen on and around your crystal's frequency for a signal. Remember that the oscillator runs all the time and, because you haven't yet connected the Morse key, your receiver is close enough to pick up the signal from the oscillator. This confirms that your oscillator is running. Switch off.

Connect the station aerial to the transmitter's aerial socket, and the receiver to the transmitter's receiver socket. Connect the Morse key, put the Receive/ Transmit–Netting switch in the receive position and switch on. You should be able to hear stations in the normal way. Now put the switch in the Transmit–Netting position. Signals in the receiver should almost disappear, as the circuit has disconnected the receiver's aerial.

Tune the receiver until you can hear your own crystal oscillator signal. This is known as *netting*, tuning your receiver and transmitter to the same frequency. Pressing the key will now transmit your signals when the switch is in the Transmit–Netting position; switch back to the receive position to listen for stations answering your call. As soon as you are happy that your circuit is functioning properly, **you must build the low-pass filter circuit** before using the transmitter regularly.

Parts list

Resistors: all 0.25 watt, 5% tolerance
R1	100 kilohms (kΩ)
R2	See text
R3	1 kilohm (kΩ)

Capacitors
C1, C2, C3	100 nanofarads (nF), or 0.1 microfarad (μF)
TC1	3–60 picofarads (pF) trimmer

Semiconductors
TR1	ZTX651
TR2	ZTX751
TR3	VN10KM

Additional items
RFC	7 turns of 32 SWG enamelled copper on a ferrite bead
Switch	Double-pole changeover (DPDT or DPCO) toggle
Crystal	For 80 m band
Crystal holder	HC25 type
Sockets	According to station fittings

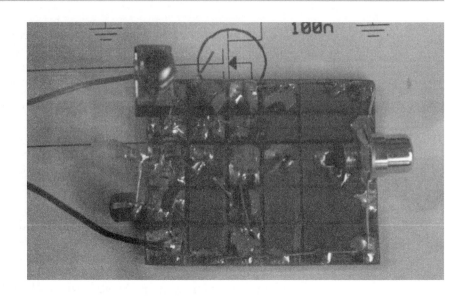

54 A 7-element low-pass filter for transmitters

Introduction

The simpler the transmitter, the more likely it is to radiate *harmonics* of its fundamental frequency. Harmonics are integral multiples of the frequency on which the transmitter is *designed* to operate. If you think you are transmitting on a frequency, *f*, for instance, you will also be radiating the harmonics of *2f*, *3f*, *4f*, . . . and so on. This results in your signals being heard on *several frequencies*, spread over a very wide frequency range. You are also contravening the terms of your licence. To avoid this, it is always advisable to use a *low-pass* filter between your transmitter and aerial. This is a filter which will *pass* your signal frequency, *f*, and all frequencies *below* it, but will *not pass* (attenuate) frequencies above *f* to any significant extent. The value of *f* is known as the *cutoff frequency* of the filter.

A design of 7-element low-pass filter

A 7-element low-pass filter (LPF) is so called because it has seven components, as the circuit diagram of **Figure 1** shows. Filters containing any odd number of elements are possible: a 3-element filter would

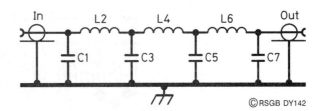

Figure 1 Circuit diagram of filter

comprise C1, L2 and C3 only, and is sometimes called a pi-network because the element disposition resembles the Greek letter pi (π); a 5-element filter would comprise C1, L2, C3, L4 and C5 only, and so on. In general, the more elements the filter has, the more effectively it attenuates signals above *f*.

The circuit of Figure 1 is designed to have an input and an output impedance of 50 Ω, which means that it can be placed in the aerial feed of any common transmitter. Filter design is a very complex business, and is best left to the experts. One such expert is W3NQN, who produced a number of computer designs of LPF using commonly available (*preferred value*) capacitors, and aimed specifically for use on amateur frequencies. The results of this work are condensed into **Table 1**. The inductors are wound on standard toroidal cores, and their details are included in the table.

Table 1 Filter component values for each brand

Band metres	C1,7 pF	C3,5 pF	L2,6 turns	L4 turns	Core type	Wire SWG
80	470	1200	25	27	T37-2	28
40	270	680	19	21	T37-2	26
30	270	560	19	20	T37-6	26
20	180	390	16	17	T37-6	24
15	82	220	12	14	T37-6	24
10	56	150	10	11	T37-6	22

Making the filter

The filter was made originally as an adjunct to the *Breadboard 80 m CW transmitter*, which you will also find in this book. It uses the same constructional technique, based on a single piece of plain, copper-clad

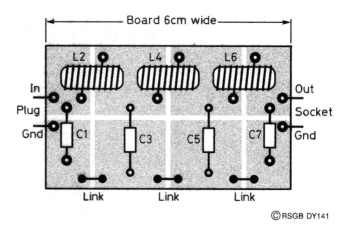

Figure 2 Layout and construction of filter

©RSGB DY141

PCB, with 'pads' created by using a sharp blade in a junior hacksaw. The cuts in the copper are shown in the layout diagram of **Figure 2**. There are two ways of mounting the components: the first way is to drill small holes in each pad, as shown in Figure 2, and mount the components through the holes in the normal PCB manner; the second way is to solder the components directly to the pads, in the way that was described for the *Breadboard transmitter*.

Winding the inductors is quite simple. All you need to remember is that each time the wire passes through the core counts as one turn. Cut off the spare wire at each end of each coil to about 1 cm, scrape off the enamel with sandpaper, and tin the exposed copper. See the transmitter description if you are unsure of how to do this. Note the wire links between each of the lower pads, forming a solid 'ground' for the elements. The prototype had a plug and socket on the ends, to match the transmitter and aerial terminations.

The type of capacitor used in the design is not critical; the polystyrene type works well.

55 Radio-frequency mixing explained

Introduction

Mixers find widespread use in electronic circuitry. Many of the projects in this book, together with every TV set and radio in the home, contain mixer circuits – a good indication of their usefulness.

Confused?

Audio mixers (as used in recording studios and radio broadcast stations) are used to **add** or 'balance' the signals from various sources such as microphones, CD players, etc. These have nothing whatsoever to do with radio-frequency (RF) mixers, and should never be confused with them.

RF mixers and beat frequencies

Instead of **adding** signals (as in the audio mixer), the RF mixer **multiples them together**. As you might expect, this has an entirely different effect. The two signals entering the mixer *beat* or *heterodyne* with each other to produce *signals on other frequencies*. One example of this occurs in sound, when two musical notes of almost the same frequency are heard together. Instead of hearing two separate notes, the listener hears one note whose *intensity* (*loudness*) appears to increase and decrease. This intensity variation is called a *beat*, and its frequency is equal to the difference in the frequencies of the two original notes. The technique is used by musicians to tune their instruments. If one note is known to be a correct frequency, the other can be tuned to it by making the beat frequency as close to zero as is possible.

Multiplying together

The process of mixing presupposes that we have a device which will automatically multiply two signals together. Fortunately, this is easy; so easy, in fact, that it often occurs when we do not want it! Multiplying is achieved by any device which is *non-linear*; this means a device whose output is not a constant factor larger than its input, something that can be achieved by many electronic devices and circuits.

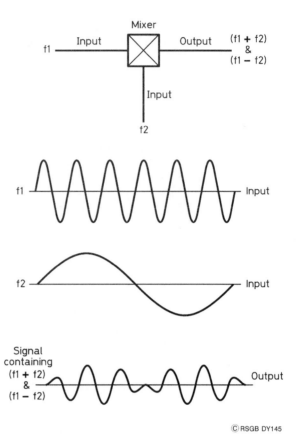

Figure 1 The effect of multiplying (or mixing) two signals together

Let us look now at what a mixer does in concrete terms. Suppose two signals, of frequencies $f1$ and $f2$ go into our mixer. These signals are shown in **Figure 1**. Putting numbers in, to make the situation clearer, suppose $f1$ is 1.000 MHz and $f2$ is 160 kHz. The beat frequency is the *difference* of these: 1.000 MHz – 0.160 MHz = 0.840 MHz, or 840 kHz. A mixer *also* produces an output at the *sum* of these frequencies; in this case the new frequency would be 1.000 MHz + 0.160 MHz = 1.160 MHz.

Suppose you fed the output of your mixer, operating with these input frequencies, into a receiver and tuned around to find what frequencies were present. You would find two signals, one at 840 kHz and one at 1.160 MHz, showing that the two 'new' frequencies were very real!

In addition to drawing out the waveform of the resultant signal, as in Figure 1, we can draw the inputs and outputs on a frequency axis, to form a *spectrum* of the signal components. This is done in **Figure 2**. The top two diagrams show the input signals at $f1$ and $f2$. The bottom diagram shows the output signals in relation to the input signals. Depending on the type of mixer used, one or both of the input signals would be removed.

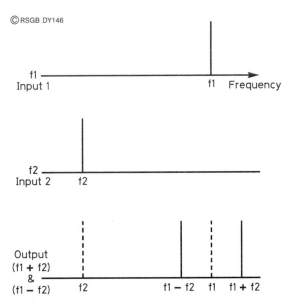

© RSGB DY146

Figure 2 The result of multiplying or mixing two signals together as seen on a spectrum analyser

A mixer in every radio

Basically, a mixer is used to change a signal from one frequency to another, something it does without altering the characteristics of the incoming signal. If the incoming signal is amplitude modulated (AM), then the frequency-changed signal would be AM also. The same applies to FM, SSB, CW, and all other modulation forms you can think of. This explains why mixers are often called *frequency changers*.

Frequency changing is the key process in the type of radio known as a *superheterodyne* (or *superhet*). By mixing the incoming signal with a variable-frequency *local oscillator* as **Figure 3**, shows, the signal can be converted to the fixed frequency of a filter and amplifier. This is useful

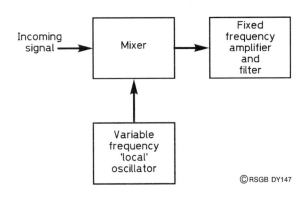

Figure 3 The basic idea of a superhet receiver

© RSGB DY147

191

because it is easier to make a very high-quality filter and amplifier at a single fixed frequency, than at a variable frequency.

All TV receivers and virtually all radio receivers (and transmitters) use mixers. Both the *Yearling* and *Colt* receivers (see the relevant projects) are superhets, and use mixers.

56 A voltage monitor for a 12 V power supply

Introduction

If for any reason, the stabilisation of your main 12 volt power supply unit (PSU) breaks down, it is possible that a voltage much higher then the nominal 13.8 V will be applied to your precious equipment. If you would like to know the instant that this occurred, and hence be able to switch things off before it was too late, then this circuit is what you need. It will give audible *and* visual indications if the voltage rises above 14.4 V, and a visual indication only if the voltage is reduced.

The circuit

The circuit uses three ICs and is shown in **Figure 1**. The circuit is powered by the PSU whose output is being monitored, and the circuit's immunity to supply line variations is secured by the 6 volt regulator, IC1. The heart of the circuit is IC2, an LM3914; it is a *bargraph driver*, which operates ten LEDs in a display resembling a thermometer – the string of lit LEDs increases in length as the voltage on pin 5 increases.

The input voltage range on pin 5 is 1.2 V maximum, making each LED correspond to one-tenth of this, which is 0.12 V, the *step size*. R1 and R2 act as a voltage divider, so that voltages of up to the maximum of 14.4 V may be applied to R1 without exceeding 1.2 V at pin 5. R3 sets the brightness of the LEDs and R4 determines the step size.

IC3 is an *opto-isolator*, a device containing an LED and a phototransistor in one package. This enables the piezoelectric sounder to operate without affecting the operation of the bargraph driver. The input to IC3 is provided by the voltage on D8, so that if any of the LEDs at or above D8 are lit, the

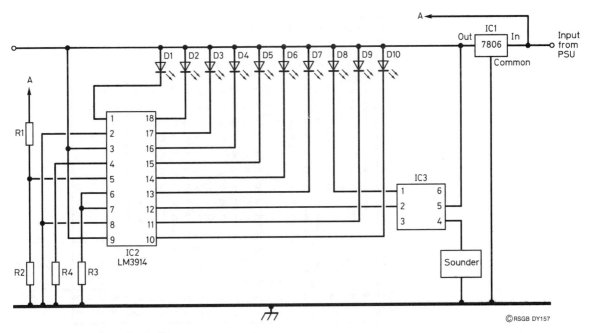

Figure 1 PSU monitor, circuit diagram

sounder will operate, indicating overvoltage. Despite the fact that some of your other equipment might be damaged by this overvoltage, the monitor circuit itself is unaffected.

Construction

The prototype was built on two pieces of Veroboard of the copper-strip variety. The main (circuit) board measured 15 strips by 25 holes, and the display (LED) board measured 4 strips by 30 holes. Track cuts are necessary in each board. The correct places are shown for the main board in **Figure 2**, but there is sufficient flexibility in the layout of the display board for a prescriptive layout not to be needed. All the anodes of the LEDs are connected to the same strip, which makes things comparatively simple!

For the main board, insert the Veropins and the wire links first, and solder them to the copper strips. Then fit the IC holders and the resistors. Fit the IC holders with their notches towards the top of the board. When fitting the voltage regulator, IC1, note than the centre lead (the 'common' lead in Figure 1) does *not* go to position A3; the track should be cut at A3, and the common lead soldered at B3. This is indicated in Figure 2.

Wire up the LED board with its ten LEDs and 11 connecting wires, each about 10 cm long. This length depends on how far away from the main

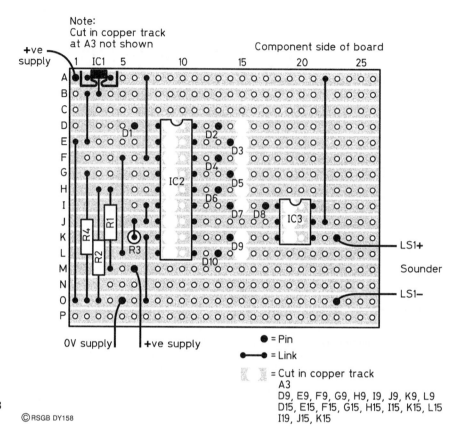

Figure 2 PSU monitor, PCB
layout

©RSGB DY158

board you are planning to mount the display. The LEDs should have
different colours: three orange, four green and three red, to indicate 'low',
'medium' and 'high' voltage.

Now solder the other ends of these leads to the Veropins on the main board,
making sure that the order is correct. Finally, connect the piezoelectric
sounder to the main board; the polarity is important, so make sure the red
lead goes to pin 4 of IC3 and the black lead to the ground rail. If you were
careful to fit the IC holders with their notches in the correct positions,
match these up with the notches on the ICs before pushing home the ICs
gently. Check your circuit for solder bridges and unwanted pieces of copper
swarf before screwing the small heat sink to the voltage regulator, IC1.

Testing

(a) **With a variable-voltage PSU.** Set the PSU for 10 V, connect the circuit
and switch on. Increase the voltage slowly, and check that each LED
lights up after the one before it. As the voltage exceeds about 14.4 V, the

first red LED should light and the sounder should operate. If an LED does not illuminate, you should immediately suspect either a dry joint or an incorrect LED polarity. The voltage at which the first red LED lights can be adjusted by varying R1; increase R1 if the LED comes on too early; decrease R1 if the LED comes on too late.

(b) **Without a variable-voltage PSU.** For your 'variable supply', you can use several AA-type 1.5 V cells (or 1.2 V NiCad cells) in series. The voltages produced by a range of cells is shown in **Table 1** – because of the lower voltage of NiCad cells, more of them are needed to produce a given voltage.

Once the operation of the circuit has been checked, it can be fitted into a plastic or metal box. Only two connections are needed for the PSU. You may want to drill some holes in the case to increase the apparent loudness of the sounder.

Table 1 Test voltages available from batteries in series

No. of batteries	Voltage, nicads	Voltage, dry cells
7	8.4	10.5
8	9.6	12
9	10.8	13.5
10	12	15
11	13.2	–
12	14.4	–
13	15.6	–

Parts list

Resistors: all 0.25 watt, 5% tolerance
R1   11 kilohms (kΩ) – see text
R2   1 kilohm (kΩ)
R3   1.2 kilohms (kΩ)
R4   18 kilohms (kΩ)

Semiconductors
IC1   L7806
IC2   LM3914
IC3   Opto-isolator – Maplin code WL35Q
D1–D3   3 mm LED, orange
D4–D7   3 mm LED, green
D8–D10   3 mm LED, red

Additional items
LS1   Piezoelectric sounder, wire leads
      6-pin DIL socket for IC3
      18-pin DIL socket for IC2
      Veroboard – two pieces for main and display boards, see text
      for sizes
      Veropins
      Heat sink for IC1
      Single-core insulated wire for links
      Insulated stranded wire for interconnecting the boards
      Case as required

57 A 1750 Hz toneburst for repeater access

Introduction

Repeaters across the UK and much of Europe need an access tone to switch the transmitter from standby ready for use. Commonly, this is a 1750 Hz tone of duration no greater than half a second. Although many UK repeaters may now be accessed using the *continuous tone-coded squelch system* (CTCSS – see the *RSGB Yearbook*), you may wish to access a repeater whose CTCSS frequency you don't know; in this case, using the universal 1750 Hz tone will gain you access. Commercial transceivers are usually fitted with an automatic toneburst, but if you are using a home-made design, then you may want to incorporate this little circuit.

Warning

This circuit uses a member of the integrated circuit family known as CMOS (complementary metal-oxide semiconductor). These use very little current and can be completed destroyed if they come into contact with the magnitudes of static electricity that most of us carry about when we walk on carpets and wear rubber shoes. You will never know if this wanton destruction has happened – all you will discover is that your circuit doesn't work and that you have tested *everything*. To avoid this problem do the following things:

1. Before you open the little packet in which the IC is supplied, touch something which you *know* to be earthed – the metalwork of any equipment which is mains earthed, for example. Then open the packet.
2. Let the IC fall gently on the bench – don't pick it out with your fingers. Touch your earthed metalwork again. Pick up the IC and insert it gently into its holder.

The circuit is safe from destruction while it is connected to the battery.

Circuit description

The circuit is shown in **Figure 1**. The tone is generated by an integrated circuit oscillator (IC1), whose frequency is controlled by a *ceramic resonator*, XL1. Its frequency is very high, and is divided down to the 1750 Hz needed by the same chip. The ceramic resonator is designed to operate at 455 kHz, the intermediate frequency of many receivers. Because all divider circuits use powers of 2, we need the oscillator to run at 448 kHz so that when it is divided by 256 ($256 = 2^8$), we end up with 1750 Hz. Try it on your calculator:

$$\frac{448\ 000}{256} = 1750.$$

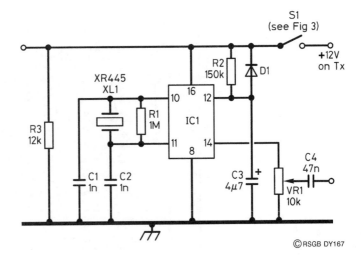

Figure 1 Toneburst module, circuit diagram

©RSGB DY167

We use C1, C2 and R1 to *pull* the frequency of the oscillator away from 455 kHz to 448 kHz. The divider chain has eight *counters* in it, and each counter divides the frequency of the signal it sees by two, giving the final division of 256.

This counting process can be stopped at any time by taking the voltage on the *reset* pin (pin 12) up to the supply voltage. When the circuit is switched on by closing S1, pin 12 is at 0 V because C3 is discharged. The oscillator runs, producing the output frequency of 1750 Hz. As time progresses, C3 charges up through R2 and the voltage on pin 12 rises. When this has risen sufficiently, and in a time determined by the values of C3 and R2, the counter resets and stays in the reset state; no division takes place and there is no output. The duration of the toneburst is thus governed by C3 and R2.

When S1 is opened, the circuit is switched off, and C3 is discharged through D1 and R3, ready for the next toneburst. If you have used a repeater, you will know that a toneburst is needed **only** to activate a repeater in the standby condition; it is not needed once a contact has been established.

VR1 adjusts the amplitude of the tone fed to the microphone, and C4 prevents any voltage that may be present on your microphone connector from damaging the integrated circuit.

Construction

The prototype circuit was built on Veroboard of the copper-strip variety, measuring 20 holes by 14 strips. The layout is shown in **Figure 2**. Make the track cuts first, and check that there are no slivers of copper wedged between adjacent tracks. Then, solder in the IC socket (with the notched end

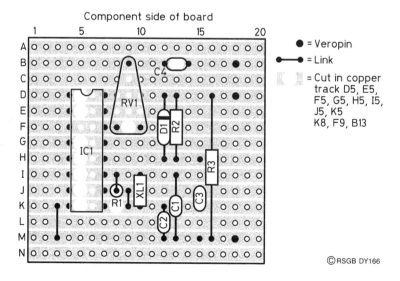

Figure 2 Veroboard component layout

©RSGB DY166

facing towards track A), the wire links and the three Veropins. Having done this, solder in the resistors, capacitors and diode, making sure that D1 and C3 are the right way round! Using your best soldering technique, solder in the ceramic resonator quickly, to prevent heat damage. Recheck your circuit, check for solder splashes and bridges, and then gently insert IC1 into its socket, matching up its notch with that of the socket.

Testing

Set VR1 to half-way and connect a crystal earpiece to the output; apply power to the circuit. You should hear the tone, lasting for about half a second. If there is no tone, disconnect your circuit from the power supply, and check for dry joints in the vicinity of pin 12. Is the diode, D1, the correct way round? Is C3 the correct way round? Did you choose to ignore the CMOS safety precautions given earlier?

Once the circuit is working, you need to decide how you are going to connect it to your transmitter. Two options are shown in **Figure 3**. If you have access to a point in your transmitter circuit that has between 9 V and 12 V positive on it during transmit, you can use this to power your circuit. As the toneburst is needed only for repeaters, the switch, S1, disconnects it when not needed, as shown in Figure 3a. If you want the circuit to be self-powered, then a 9 volt PP3 battery may be used; Figure 3b shows this configuration.

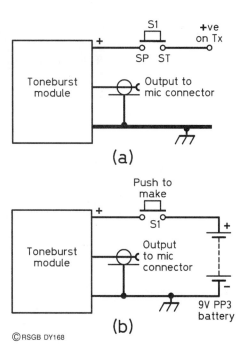

Figure 3 Toneburst module: two alternative switching arrangements

©RSGB DY168

The output from the circuit board is fed directly into the microphone socket, in parallel with the microphone itself; use thin coaxial or screened cable for this lead, or you may induce hum into the microphone circuit and suffer from RF breakthrough into the audio circuits. To adjust the setting of VR1, start with it at the zero output position and connect a dummy load to your transmitter. Slowly, increase the output while monitoring your transmitted signal on another nearby receiver. Make sure you do **not** increase the output so far that the signal sounds distorted. If you would prefer that the tone was on continuously while you made this adjustment, simply connect a wire across C3 remembering, of course, to remove it as soon as you have completed the test!

Parts list

Resistors: all 0.25 watt, 10% tolerance (or better)

R1	1 megohm (MΩ)
R2	150 kilohms (kΩ)
R3	12 kilohms (kΩ)
VR1	10 kilohms (kΩ) horizontal preset

Capacitors: all 16 V WKG or higher

C1, C2	1 nanofarad (nF) disc ceramic
C3	4.7 microfarad (μF) tantalum bead
C4	47 nanofarads (nF) disc ceramic

Semiconductors
    IC1   4060
    D1   1N4148

Additional items
    XL1   XR455
           Veroboard (see text for size)
           Veropins (3)
    S1   Switch (momentary action push-to-make SPST)
           16-pin DIL socket for IC1
           Single-core insulated wire for links
           Coaxial or screened cable for microphone connection

58 A circuit for flashing LEDs

Introduction

There are many occasions when one's attention needs drawing to the fact that something important has happened. A single red light coming on is seldom sufficient to attract attention, particularly if it is surrounded by other lights and indicators. The eye is known to be very sensitive to *changes* in its peripheral vision; such changes can be brought about by movement or by differences in light level – a flashing light, for example. So, a circuit that flashes a single LED or a pair of LEDs finds plenty of uses in the amateur station.

Warning

This circuit uses a member of the integrated circuit family known as CMOS (complementary metal-oxide semiconductor). These use very little current and can be completed destroyed if they come into contact with the magnitudes of static electricity that most of us carry about when we walk on carpets and wear rubber shoes. You will never know if this wanton destruction has happened – all you will discover is that your circuit doesn't work and that you have tested *everything*. To avoid this problem do the following things:

1. Before you open the little packet in which the IC is supplied, touch something which you *know* to be earthed – the metalwork of any equipment which is mains earthed, for example. Then open the packet.
2. Let the IC fall gently on the bench – don't pick it out with your fingers. Touch your earthed metalwork again. Pick up the IC and insert it gently into its holder.

The circuit is safe from destruction while it is connected to the battery. However, when the battery is removed, the same care should be exercised with its handling, because there is no supply decoupling capacitor across the IC.

Basic description

LEDs can be made to flash (switch on and off) by driving them from sources that switch on and off. Such a source is an *astable multivibrator*. If you have built or read about *A basic continuity tester*, elsewhere in this book, you will have come across such a beast before. That circuit used an astable multivibrator made from two transistors. This new circuit achieves the same behaviour from a single integrated circuit, the CMOS 4011. To give it its full description, the 4011 is a *quad 2-input NAND gate*. Quite a mouthful, but all it means is that inside the chip are four NAND gates, each with two inputs.

A NAND gate needs a positive voltage (known as a *logic 1*) on *both* inputs in order to produce zero volts (known as *logic 0*) at the output. Two NAND gates can be connected, as are A and B in **Figure 1**, to make our astable multivibrator. The combination of A and B has been described as the most perverse circuit in electronics; as soon as the output goes to logic 1, the circuit decides that it would prefer to have a logic 0 there, and switches over. With logic 0 at the output, the circuit now prefers to have logic 1 there, and so it goes on! We are going to use this continuous switching backwards and forwards to flash two LEDs. The rate at which A and B 'change their minds' is the frequency at which our LEDs will flash, and is controlled by the charging and discharging times of C1 through R2 and C2 through R1. As the values of R1 and R2 are the same, and those of C1 and C2 are the same, the ON and OFF states of the circuit are the same.

Gates C and D do not contribute to the flashing action; they act as *buffers* to isolate the LEDs from the multivibrator circuit itself. You will find in electronics that an oscillator is seldom used to drive another device *directly*; there is usually a buffer between it and the stage it drives.

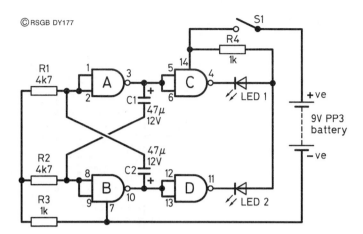

©RSGB DY177

Figure 1 Flashing LEDs, circuit diagram

A characteristic of all multivibrators, astable or not, is that they have two outputs. In this case, those outputs are at pins 3 and 10, which are then buffered and appear at pins 4 and 11, respectively. When one output is at logic 1, the other is at logic 0, and vice versa. This means that LED1 is off when LED2 is on, and LED1 is on when LED2 is off, the two states switching backwards and forwards at the frequency of the oscillator.

Construction

Read the warning at the beginning of this article again. It is not intended to scare you off from building this, but is a genuine piece of advice which can save you time and irritation when all your labours result in a circuit that doesn't work! That extra bit of care can make all the difference!

Veroboard (the copper strip type) is used for the layout, shown in **Figure 2**. It measures 20 holes by 12 strips. **Be aware that there is no row 'I' in the layout, so don't miscount when you are placing components on the board!**

Firstly, cut the tracks using a 3 mm (⅛ inch) twist drill held between thumb and forefinger; check that there are no slivers of copper bridging any of the tracks, and that the tracks have been completely cut by the drill. Solder in the components carefully. Leave the IC in its carrier for the time being, and solder in the IC socket, with the notched end facing row A. On completion, check the circuit carefully. If you are happy that it is correct, follow the instructions given earlier and fit IC1 into its socket, matching up the two notches. Connect the battery and switch on. The two LEDs should flash on and off alternately. If only one LED flashes, you have probably connected the other one the wrong way round. Switch off, check and correct if necessary. If neither LED flashes, you must have a significant error in your circuit, which will need checking *again*! Or did you choose to ignore the handling precautions for the CMOS chip?

Figure 2 Flashing LEDs, component layout

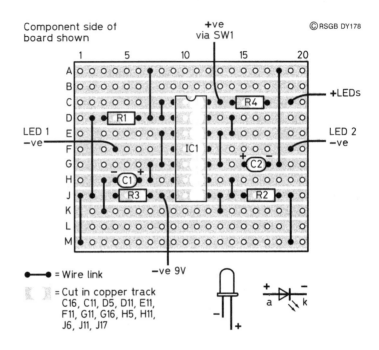

The board can be mounted near to the point where you want your flashing LEDs to be seen, although long leads to the LEDs are acceptable. The LEDs can be different colours – it's all up to you now!

Parts list

Resistors: all 0.25 watt, 10% tolerance or better
R1, R2	4700 ohms (Ω)
R3, R4	1 kilohm (kΩ)

Capacitors
C1, C2	47 microfarads (μF) 12 V WKG

Semiconductors
IC1	4011

Additional items
LED1, LED2	Any size of LED, any colour
S1	SPST on/off
	Plastic box if needed, 8.5 by 5 by 2.5 cm

Source

Components are available from Maplin.

59 Digital logic circuits

Introduction

Logic circuits form the backbone of even the most advanced computer, yet their basic operation can be demonstrated by a couple of switches, a battery and a bulb.

Logic using switches

Everyone reading this article will look at **Figure 1** and know immediately how it works and be able to write down something like 'When switch A and switch B are closed, the light will come on'. Without knowing it, you have written down a logic statement involving the so-called **AND** operation; the light comes on only when switches A **AND** B are ON. Below the circuit in Figure 1, is a table showing the only possible positions of the two switches and the state of the bulb for each position. This is called a *truth table*, and is frequently used in logic analysis.

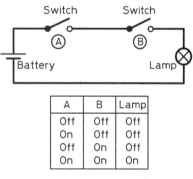

A	B	Lamp
Off	Off	Off
On	Off	Off
Off	On	Off
On	On	On

Figure 1 Switches and lamp AND gate

©RSGB DY182

Figure 2 shows a different circuit. Here, the two switches are in parallel rather than in series, as was the case in Figure 1. Again, if you analyse the circuit in words, you would say that the light will be on when switch A **OR** switch B is ON. This is an example of the **OR** operation, and its truth table is shown in Figure 2. The statement above is not complete, however; can you see why? The truth table will show you. The light comes on if A is ON, **OR** if B is ON, *OR* if A **AND** B are both ON. That third condition is easy to miss, but don't worry about it!

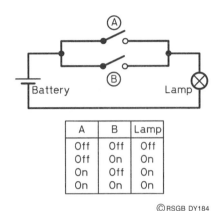

A	B	Lamp
Off	Off	Off
Off	On	On
On	Off	On
On	On	On

Figure 2 Switches and lamp
OR gate

©RSGB DY184

Believe it or not, some very complicated logic is possible (in theory) using switches and lights, but it is highly impractical and would be very slow. This is where electronic logic circuits come in.

Switches with no moving parts

You may have come across projects in this book where a statement is made such as '. . . the transistor is being used as a switch . . .'. Transistors *can* be used as switches, as were thermionic valves in the world's first pro-grammable computer *Colossus*, at Bletchley Park. However, technology has moved on from valves, through transistors to *logic gates*, combinations of electronic switches designed specifically to perform logic functions.

These act on voltage levels as their inputs and produce changes in voltage levels as their outputs. A positive voltage is called *logic 1*, and corresponds to a switch being ON in our previous descriptions; a zero voltage is called *logic 0*, and corresponds to a switch being OFF. The output from a logic gate (normally labelled Q) is also logic 1 or logic 0, corresponding to our light being ON or OFF, respectively, in our switch analogy.

Many logic devices operate from a stabilised 5 V supply, and this determines the *ideal* voltages corresponding to the two logic states:

logic 0 ≡ 0 V,

logic 1 ≡ 5 V.

The world isn't an ideal place, so the *real* voltage ranges used by the logic gates are:

logic 0 ≡ 0.0 to 0.4 V,

logic 1 ≡ 3.0 to 5.0 V.

The **AND** circuit of Figure 1 is now called an **AND gate**, and requires logic 1 inputs on A **AND** B to produce a logic 1 at the output. **Figure 3** shows this. The truth table is identical with that of Figure 1 – logic 1 replaces ON and logic 0 replaces OFF. Now compare Figure 2 with **Figure 4** – circuits and truth tables for the **OR** function. Again, we have exact similarity.

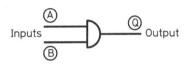

Inputs		Output
A	B	Q
0	0	0
0	1	0
1	0	0
1	1	1

©RSGB DY183

Figure 3 Electronic AND gate

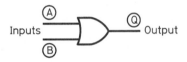

Inputs		Output
A	B	Q
0	0	0
0	1	1
1	0	1
1	1	1

©RSGB DY190

Figure 4 Electronic OR gate

There is another very common logic gate, which performs the **NOT** function. It is easy to understand. Just ask yourself the question 'What is **NOT** logic 0?', and the answer is obviously 'logic 1'. Similarly, logic 0 is **NOT** logic 1. A **NOT** gate simply changes the logic state of the input; it is also known (because of this behaviour) as an *inverter*. Its symbol and truth table can be found in **Fig 5**. Note the little circle on the output of the gate in Figure 5. In logic circuits, this symbol always implies inversion, or the presence of a **NOT** gate. Keep an eye open for it!

So far, the logic functions we have discussed have all been words which we use in everyday language, which has made the electronic interpretation of them relatively easy. Now we must introduce a function for which there is *no* analogy in normal speech – the **NAND** function. This means a

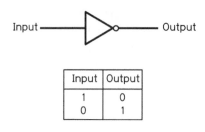

Input	Output
1	0
0	1

Figure 5 Electronic NOT gate

©RSGB DY185

combination of an **AND** gate and a **NOT** gate, and the sharp-eyed reader will have spotted the little circle added on to the normal **AND** gate symbol in **Figure 6**!

To make things easier to understand, the truth table in Figure 6 has four columns, not three as in previous tables. The third column is the standard **AND** output – compare it with the third column in Figure 3. That is the output from the **AND** gate *before* it encounters the little circle that inverts it, so the final output from the **NAND** gate is the output of the **AND** gate, inverted! The third and fourth columns are the inverse of each other.

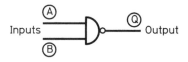

Inputs		and	Output
A	B		Q
0	0	0	1
0	1	0	1
1	0	0	1
1	1	1	0

Figure 6 Electronic NAND gate

©RSGB DY186

Inputs		or	Output
A	B		Q
0	0	0	1
0	1	1	0
1	0	1	0
1	1	1	0

Figure 7 Electronic NOR gate

©RSGB DY187

We can add the inverting operation to the output of an **OR** gate also, producing a **NOR** gate! The symbol and its truth table are shown in **Figure 7**. Again, notice that the final output is that of the ordinary **OR** gate, inverted!

These new functions of **NAND** and **NOR** are used more than the **AND** and **OR** functions because it makes other circuits easier to design using combinations of these gates.

A taste of Boolean algebra

The design of circuits using combinations of logic gates usually begins with a little mathematics, where the functions to be implemented are analysed. The mathematics used is surprisingly simple, and is a slightly changed version of ordinary algebra called *Boolean algebra*, which allows manipulation of logic functions to be made. Normal algebra has operations in it such as addition, subtraction and multiplication and division. The mathematician Boole found that the logical **AND** operation could be handled by the algebraic operation of *multiplication* (symbols $\times$ or $\bullet$), and the **OR** operation by the algebraic operation of *addition* (symbol +). The **NOT** operation involved a new symbol, that of a bar over the input being inverted, such as $\overline{A}$.

So, our five basic logic operators can now be written in a mathematical form:

AND	$Q = A \times B$
OR	$Q = A + B$
NOT	$Q = \overline{A}$
NAND	$Q = \overline{A \times B}$
NOR	$Q = \overline{A + B}.$

Using logic operations in a mathematical form enables the most complex logic to be designed, simplified and converted into circuit diagram form in a very efficient and rapid way. There are more logic operators than the five we have considered here but, in general, they can all be broken down into combinations of these gates alone!

60 A resistive SWR indicator

Introduction

When a transmitter produces some output power, we want to make sure that as much as possible of this power is radiated by the aerial. This often requires the use of an *aerial tuning unit* (ATU), which matches the aerial impedance to that of the transmitter.

How do you know when this matching has been achieved? The most usual way is to use a *standing-wave ratio* (SWR) indicator. If the impedance of the aerial does not match that of the transmitter output, some of your transmitter power (also known as the *forward* power) is reflected back along the aerial feeder and back into the transmitter, where it causes excess heating. The forward and reflected waves interact along the feeder to produce a wave whose position remains constant, and which is therefore called a *standing wave* or a *stationary wave*. An SWR meter simply indicates forward power and reflected power, and adjustments are made to your ATU until the reflected power is as small as possible (ideally zero, of course). If there is no reflected power then, by a process of elimination, all your forward power is reaching the aerial!

Sampling the RF

Whatever type of SWR indicator you use, it must use some sort of sampling circuit to pick up the forward and reflected waves. The project *A standing-wave indicator for HF*, elsewhere in this book, uses a toroidal transformer to separate the readings for the forward and reverse waves. This design differs in that it measures the voltages across resistors through which the RF current is passing. Its advantages are:

(a) it uses cheap parts – four resistors, two capacitors and a diode, together with a rotary switch, a surplus meter, a preset potentiometer and two sockets;

(b) within this SWR indicator, there is always a resistive path for the RF current from the transmitter, formed by R1, R2 and R3; this can prevent damage to simpler home-made transmitters, which may be damaged during adjustment of the ATU when using more conventional SWR indicators.

The only disadvantage of this form of indicator is that it must be switched out of circuit once the ATU has been adjusted for a particular band.

Construction

The SWR indicator is very simple to build, as most of the components can be mounted on the back of the 3-way rotary switch. This is shown in **Figure 1**. The switch is a 4-pole, 3-way rotary type, of which only two poles are used.

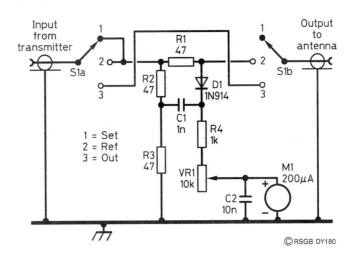

Figure 1 Resistive SWR meter, circuit diagram

Because the other switch contacts are not being used, they can be employed as support tags for other components. The ground wires are all soldered on to the metal frame of the switch. If your switch frame is of all-plastic construction, then a 12 SWG copper wire run around the switch will make a good earth connection to the metal case for the leads shown in Figure 1. The preset potentiometer used to control the sensitivity of the circuit can be mounted directly on the meter tag.

Resistors R1, R2 and R3 handle the RF power during the tuning-up process. If you have them, use 1 watt resistors; otherwise, you can use two 100 ohm half-watt resistors in parallel for each of R1, R2 and R3. The meter, M1, can be any DC type of sensitivity around 200 μA.

In use

First, find a clear frequency, and without the indicator in circuit, check that the frequency really *is* clear by asking and identifying yourself. If it is, connect the indicator between the transmitter and the ATU which, in turn, is connected to your aerial. Turn SW1 to the SET position and key the transmitter. Adjust VR1 until the meter reads full scale. Switch off the transmitter. Turn SW1 to the REF position and key the transmitter again. Adjust the ATU until the lowest reading is obtained on the meter. Switch off the transmitter. For the chosen frequency, you have adjusted your ATU for minimum reflected power and hence the lowest SWR. You will need to repeat the process when you change bands, and possibly when you change frequency within the same band. Switching SW1 to the OUT position, you are ready to transmit. You may have noticed that it is good practice to switch the transmitter off when operating SW1. Get into that habit!

Parts list

Resistors
R1, R2, R3	47 ohm (Ω) 1 watt (or 2 × 100 Ω, $\frac{1}{2}$ watt, see text)
R4	1 kilohm (kΩ) 0.25-watt
VR1	10 kilohms (kΩ) preset

Capacitors
C1	1 nanofarad (nF)
C2	10 nanofarads (nF)

Semiconductors
D1	1N914 or similar

Additional items
SW1	4-pole 3-way rotary switch, of which only 2 poles are used
M1	200 μA DC
	Case

61 An audio filter for CW

Introduction

This is a simple passive circuit (it has no power supply) that adds some audio selectivity for Morse code reception and also includes a very simple noise limiter that gives a visual indication of when noise spikes are being removed!

The circuit

Figure 1 shows the complete circuit. The tuned circuit of C1 and L1 resonates very close to 800 Hz, so initially you will have to tune a signal in carefully until it sounds loudest – you will soon be able to do this without thinking. The two LEDs connected back to back across the signal path act as a noise limiter, reducing the amplitudes of static crashes and noise from car ignition systems, etc. The LEDs blink when they conduct – this is not necessary to the operation of the circuit, but adds a little colour to your listening! The noise limiter *does* make listening more comfortable, though.

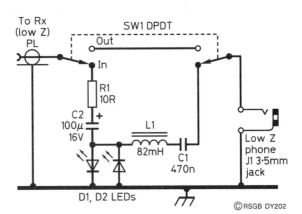

Figure 1 CW filter, circuit diagram

Construction

The circuit layout is shown in **Figure 2**. Point-to-point wiring is used, with a small tag-strip being the only item used for the extra support of C2 and R1. The LEDs and SW1 support the other components. An aluminium box (**Figure 3**) is used to make the circuit tidy and usable.

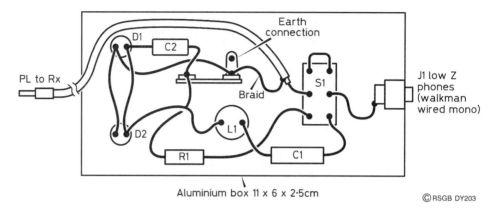

Figure 2 CW filter, component layout

Aluminium box 11 x 6 x 2·5cm

©RSGB DY203

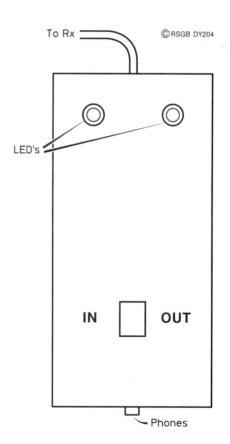

Figure 3 CW filter, front panel

You must use low-impedance headphones for the circuit to perform properly. Plug the completed unit into the headphone socket of your receiver and adjust the volume so that the LEDs are just *not* blinking on normal audio. The unit should be switched out of circuit for speech listening; this is the purpose of the toggle switch, SW1.

Parts list

Resistor
R1   10 ohms (Ω), 0.25 watt, 10% tolerance

Capacitors
C1   470 nanofarads (nF)
C2   100 microfarads (μF) electrolytic, 16 V WKG

Inductor
L1   82 millihenries (mH)

Semiconductors
D1, D2   LEDs

Additional items
SW1   DPDT
J1   3.5 mm jack socket
PL   Jack plug to suit receiver
                  Aluminium box, approx. 11 by 6 by 2.5 cm

Source

Components are available from Maplin.

62 An electronic die

Introduction

Throwing a die is a venerable way of generating a 'random' number between 1 and 6. A die is easy to lose, so here is an electronic die-throwing circuit which brings the technique up to date and serves as another application of logic circuits. If you need a reminder of the basics of logic, refer to *Digital logic circuits*.

Warning

This circuit uses members of the integrated circuit family known as CMOS (complementary metal-oxide semiconductor). These use very little current and can be completed destroyed if they come into contact with the magnitudes of static electricity that most of us carry about when we walk on carpets and wear rubber shoes. You will never know if this wanton destruction has happened – all you will discover is that your circuit doesn't work and that you have tested *everything*. To avoid this problem do the following things:

1. Before you open the little packet in which the IC is supplied, touch something which you *know* to be earthed – the metalwork of any equipment which is mains-earthed, for example. Then open the packet.
2. Let the IC fall gently on the bench – don't pick it out with your fingers. Touch your earthed metalwork again. Pick up the IC and insert it gently into its holder.

The circuit is safe from damage while it is connected to the battery.

Description

The circuit is shown in **Figure 1**. IC1a and IC1b form an *astable multivibrator*, similar to that used in 'An LED Flasher' also in this book. It runs constantly, and its output is fed to IC2 via a single NAND gate, IC1c. The other input of IC1c is held at 0 V by R3 and C3. Whenever one input of a NAND gate is logic 0 (or 0 V), there can be *no output* from the gate, irrespective of what is happening at other inputs. So, despite the fact that the oscillator (or *clock*) is running all the time, its square-wave output never

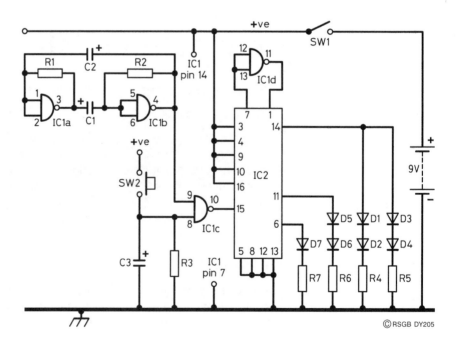

Figure 1 Electronic dice, circuit diagram

©RSGB DY205

reaches IC2 until the 'other' input of IC1c is brought away from logic 0. This happens when SW2 is pressed. Pressing SW2 momentarily, as is normally done, charges up C3 to 9 V, putting a logic 1 on the second input of IC1c and allowing the clock signal through to IC2. As the switch is released, C3 begins to discharge through R3, gradually lowering the voltage on pin 8 of IC1c. As this voltage crosses 4.5 V (half the supply voltage), IC1c then treats this input as logic 0, which cuts off the clock signal from IC2 again. Thus, when SW2 is momentarily pressed, the clock signal is fed to IC2 for a short interval of time, before being blocked again.

IC2 is a binary counter, and the outputs that concern us are from pins 14, 11 and 6. Pin 14 is the *most significant bit* and pin 6 the *least significant bit*. See the panel for an explanation of what happens here.

To understand these outputs, you should be able to count in binary, using three bits. At the start of the counting process, all the bits have zero values, i.e. 000. The left-most bit is called the *most significant bit*, and the right-most bit is called the *least significant bit*. (When we write numbers normally, a number 1 in the left-hand position represents one hundred, 1 in the middle column represents ten, while the number 1 in the right-hand position means one. One hundred is a more significant number than one, hence the nomenclature.)

The three bits of our binary number have values, from left to right, of 4 (=2^2), 2 (=2^1) and 1 (=2^0). This means that a number 1 in the left-hand column signifies the normal number four, in the middle column it would represent two, and in the right-hand column, it would represent one. This should help you understand the patterns of bits which emerge as the clock waveform is counted, as the next paragraph explains.

As each cycle of the clock enters IC2, it increments its internal counter and the values of that counter are shown by the states of pins 14, 11 and 6. After the first pulse, these three pins would have states corresponding to 001; after the second, 010; after the third, 011; after the fourth, 100; after the fifth, 101; after the sixth, 110. This sequence of 3-bit numbers represents a binary count from 1 to 6 in 'normal' parlance. These six states are used to illuminate the conventional pattern of dots on a die, using LEDs.

As the clock cycles are counted, the LEDs flicker as the die is 'rolled'. Resistors R4 to R7 are used to limit the current through the LEDs. When the counter stops, it can be in any of the positions shown in **Table 1**. Because of

Table 1 Dice number relative to counter output

Step	Pins			Dice number
	14	11	6	
1	0	0	1	4
2	0	1	0	2
3	0	1	1	6
4	1	0	0	1
5	1	0	1	5
6	1	1	0	3

the way in which the dots are grouped on the faces of a die, the wiring of the LEDs is simpler than it might otherwise be. You can see from Figure 1 that there are really only three sets of connections to the LEDs – one from each bit of the counter output. When pin 14 is at logic 1, four LEDs are lit, corresponding to the number four. For the number five, pins 14 and 6 will be at logic 1. For the number one, only pin 6 is at logic 1. For three, pins 11 and 6 are at logic 1, for six, pins 14 and 11 are at logic 1, and for two, only pin 11 is at logic 1. These conditions are summarised in Table 1.

Construction

Two pieces of Veroboard (of the strip type) are needed – one for the main circuit (**Figure 2**) and the other for the display LEDs (**Figure 3**). The display board is easier to build, so we will do that first. It measures 20 holes by 18 strips.

Cut the tracks as shown in Figure 3, using a 3 mm ($\frac{1}{8}$inch) twist drill rotated between thumb and forefinger. Insert and solder the Veropins, resistors and the wire links. Then solder the LEDs in place, making sure their polarities are correct. The LEDs are mounted proud of the board (**Figure 4**, which will help when you come to fix the board into a case.

The main board measures 32 holes by 18 strips. As before, remove the tracks in the places shown in Figure 2. Note that track 'L' is *not* broken under the position for IC2. Solder in the IC holders and the links. Solder in R1 and R2 vertically, and R3 horizontally. Then solder in C1, C2 and C3; these all being electrolytics, check their polarities. Solder in the Veropins and prepare the connections to those components not on the main board, using stranded, insulated wire. Check the circuit carefully, and read the handling precautions given earlier before carefully inserting the two ICs.

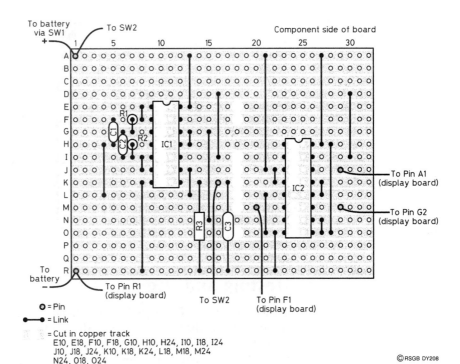

To battery via SW1

To SW2

Component side of board

To Pin A1 (display board)

To Pin G2 (display board)

To battery −

To Pin R1 (display board)

To SW2

To Pin F1 (display board)

○ = Pin

●—● = Link

= Cut in copper track
E10, E18, F10, F18, G10, H10, H24, I10, I18, I24
J10, J18, J24, K10, K18, K24, L18, M18, M24
N24, O18, O24

Figure 2 Main board, component layout

©RSGB DY208

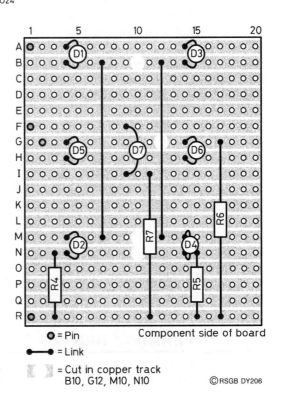

○ = Pin

●—● = Link

= Cut in copper track
B10, G12, M10, N10

Component side of board

Figure 3 Board for the six LEDs

©RSGB DY206

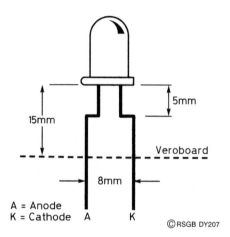

Figure 4 D7 lead bending

Testing and use

When the circuit is switched on, it is likely that the centre LED only will be lit. Press SW2, the 'throw' switch. All the LEDs should flicker, even after the switch is released, but only for a second or so. Then a standard pattern should show, signifying numbers between one and six. If this does not happen, switch off and check your circuit again. Are there any missing links? Are there any obvious dry joints (usually dull instead of shiny)? Can you see any solder bridges between tracks? Did you choose to ignore the handling precautions? If none of these results in a working circuit, try some fault-finding.

● Disconnect the lead to D7 on the main board (marked 'to pin F1' on Figure 2). Touch the lead on to the positive supply rail. D7 should light. Now transfer this lead on to pin 4 of IC1. If it flickers, then the oscillator is running (which it should).
● Transfer this lead to pin 10 of IC2. D7 should flicker when SW2 is pressed, but should be on permanently when it is released. This means that the correct signals are reaching IC2 from IC1.
● If one of the chains of LEDs (i.e. D1 and D2, D3 and D4, D5 and D6) does not light, you may have one *or both* LEDs the wrong way round.

When the die is working, you may care to experiment with the values of R1 and R2, but you should keep their values the same, i.e. R1 = R2. The larger the values of these, the more slowly the die will appear to 'roll'.

Finishing touches

On completion of the project, you will want to mount it in a smart case; any plastic box is suitable for this, with the display board mounted so that the

LEDs protrude through the top and are fixed to it using LED clips. The way in which you mount both boards to the case is entirely up to you.

Parts list

Resistors: all 0.25 watt, 10% tolerance, or better
R1, R2	5600 ohms (Ω)
R3	15 kilohms (kΩ)
R4–R7	470 ohms (Ω)

Capacitors
C1, C2	1 microfarad (μF) electrolytic, 16 V WKG
C3	68 microfarads (μF) electrolytic, 16 V WKG

Semiconductors
IC1	4011
IC2	4029
D1–D7	LEDs, any size and colour

Additional items
SW1	On-off switch SPST toggle
SW1	Push-to-make, non-latching
	Veroboard, 2 pieces, see text for sizes
	Veropins, 10
	Stranded insulated wire for general wiring
	Single-core insulated wire for links
	PP3 battery clip
	PP3 battery
	Plastic box to suit
	LED mounting clips (7)
	Means of mounting boards to box

Source

Components are available from Maplin.

63 The absorption wavemeter

Introduction

The purpose of the absorption wavemeter is to check that a transmitter is radiating within the correct waveband, and to detect any spurious *harmonic* emissions. A harmonic frequency is an integral multiple of the carrier frequency – e.g. if the carrier is at a frequency f, the harmonics are at frequencies $2f$, $3f$, $4f$ . . . and so on.

It can also be employed as a relative field-strength indicator, being used in experimentation with aerials. What it *cannot* do is to measure the transmitted frequency accurately; all it can do is to place the signal within a particular band, say the 7 MHz or the 18 MHz band (or their harmonics).

How it works

You may recognise the circuit of **Figure 1** as a crystal set with a meter replacing the headphones. It is simply the parallel combination of an inductance with a variable capacitance, which constitutes a resonant circuit with a variable frequency. A dial on the variable capacitor can be roughly calibrated with frequency. The fixed capacitor across the meter helps to improve the meter reading by making it read the peak value of the carrier wave rather than the average value.

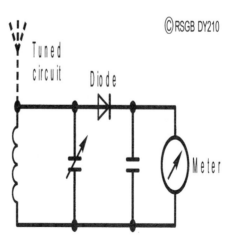

©RSGB DY210

Figure 1 Absorption wavemeter circuit

When the wavemeter is situated in the vicinity of a relatively strong radio-frequency (RF) field, the inductor absorbs a small amount of energy from that field. If the circuit is resonant at the same frequency as that of the transmitter, an RF voltage is produced across the coil and is proportional to the strength of the RF field. The RF voltage is *rectified*, or *detected* (turned from AC into DC) by the diode and the second capacitor, and is displayed by the meter.

The tuned circuit

The formula for calculating the resonant frequency of a coil and capacitor is relatively simple. It is

$$f = \frac{1}{2\pi\sqrt{LC}},$$

where f is the resonant frequency (Hz),
    L is the inductance (H), and
    C is the capacitance (F).

If this formula puts you off, there is this slightly easier version:

$$f = \frac{159}{\sqrt{LC}},$$

where f is the frequency in **MHz**,
    L is the inductance in **microhenries** (μH), and
    C is the capacitance in **picofarads** (pF).

When a variable capacitor is specified, it is usual to quote its maximum capacitance, i.e. the value when the vanes are fully meshed. When the value of the capacitor is small, it is usual to quote its minimum capacitance also, the value when the vanes are fully open. If you use a variable capacitor of 10–100 pF with a 100 μH inductor, you will use the value of 10 pF to calculate the maximum resonant frequency, and the value of 100 pF to calculate the minimum resonant frequency, giving a tuning range of 1.6 MHz to 5 MHz. Check it yourself!

64 An HF absorption wavemeter

Introduction

An absorption wavemeter is basically a simple tunable detector circuit, such as would be used in a crystal set, but with the headphones replaced by a meter, in order to indicate the strength of the received signal.

The circuit

Figure 1 shows the circuit of the wavemeter. It is designed to cover all the HF bands from 1.8 MHz to 28 MHz in four switchable ranges. There is no built-in method of amplification, so a sensitive meter is needed in order to indicate sufficiently using the available absorbed energy. The meter does not need calibrating, it serves to produce only an indication of the absorbed power, *not* its absolute value.

An absorption wavemeter does not have outstanding selectivity, mainly because of the loading effect of the meter. This problem is ameliorated here (but only slightly) by the inclusion of R1 in series with the meter. It gives

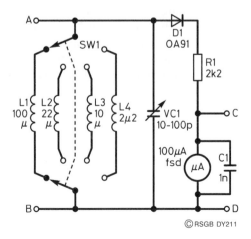

Figure 1 HF absorption wavemeter, circuit diagram

© RSGB DY211

improved selectivity at the expense of a slight loss of sensitivity. The appearance of capacitor C1 across the meter is discussed in *The Absorption Wavemeter*, in the previous section.

Coils

These can be purchased (see the parts list) or hand-wound on short lengths of PVC water pipe or conduit, and the coil-winding details are given in **Table 1**. In theory, the 100 pF tuning capacitor would cover all the HF bands with only three coils, omitting the 22 μH coil. Sample calculations are explained in the previous article. However, its inclusion avoids the common problem of having some bands cramped at the extreme edges of the tuning scales.

Table 1 Close-wound coil values

L (μH)	No of turns	Span (mm)	SWG
PVC FORMER OF OUTSIDE DIAMETER 25 mm			
2.2	9	9	20
10	24	24	20
22	44	44	20
100	110	60	26
PVC FORMER OF OUTSIDE DIAMETER 20 mm			
2.2	10	10	20
2.2	9	5	26
10	32	32	20
10	24	12	26
22	63	62	20
22	41	21	26
100	134	70	26
PVC FORMER OF OUTSIDE DIAMETER 19 mm			
2.2	11	11	20
2.2	9	5	26
10	35	34	20
10	25	13	26
22	68	67	20
22	44	22	26
100	157	79	26

Construction

Hand-wound coils, using 19–25 mm PVC formers, have single-layer windings, with their ends secured by threading the wires through two small holes drilled in the formers at each end of each winding.

You may have wondered about the use of the two-pole switch SW1, as all the lower ends of the coils are grounded anyway. The answer lies in the ease of construction. Each coil can be mounted on the tags of the switch for support. If you are using the commercial coils, they are small enough to be accommodated on a miniature rotary switch; a larger rotary switch is required for hand-wound coils. The specification of a 6-way switch allows for the addition of extra frequency ranges at little extra cost. Rotary switches such as the 2-pole, 6-way variety have an adjustable end-stop which allows the number of ways to be set anywhere between 2-ways and 6-ways. Remove the nuts from the switch, and a 'washer' will fall out. It is not a washer, really, and has a little piece of metal bent over at right angles to the washer. This fits into one of 12 holes in the body of the switch, and prevents the shaft from turning through 360° and selecting the number of ways.

Calibration

This process needs a commercial amateur-bands transceiver, if it is to be done quickly and accurately.

- Connect the transceiver to a dummy load, and connect a piece of flexible wire from point A in Figure 1 and wrap two or three turns around the cable between the transceiver and the dummy load.
- Turn the transceiver power level control to minimum, and set the frequency at 1.81 MHz and the mode to FM or AM. Switch to transmit.
- With the wavemeter set to its lowest frequency range (i.e. 1.8–3.5 MHz), rotate the tuning capacitor until maximum deflection is obtained from the meter needle. Write the frequency on the inner ring of the wavemeter scale at that point.
- Switch off and retune the transmitter to 3.5 MHz. Repeat the above process.
- Repeat, using switch range 2 for 3.5, 7 and 10 MHz, range 3 for 7, 10 and 14 MHz, and range 4 for 14, 18, 21, 24 and 28 MHz.

An alternative source of calibration signal could be an HF signal generator, with a single-turn loop of wire at the remote end of its cable, and the wire from the wavemeter brought close to the loop.

Remember that the marks on the scale represent frequency *bands*, not precise frequencies.

Extending the range

If you would like to experiment with increasing the range to the lower VHF band, then 50 MHz should be achievable with an additional self-supporting coil (no former). Try three turns of 20 SWG enamelled copper wire of 25 mm ID (internal diameter). Wind it on some 25 mm PVC pipe, as before, then slide out the pipe! The same result could be achieved with four turns of 19 or 20 mm ID, or 12 turns of 7.1 mm ID.

Parts list

Resistor
  R1   2200 ohms (Ω) 0.25 watt, 10% tolerance

Capacitors
  C1   1 nanofarad (nF) min. ceramic
  VC1   100 picofarad (pF) variable

Inductors
  L1   100 microhenries (μH)
  L2   22 microhenries (μH)
  L3   10 microhenries (μH)
  L4   2.2 microhenries (μH)

Semiconductors
  D1   Germanium OA91 or 1N4148

Additional items
  SW1   Rotary 2-pole, 6-way
            Moving-coil meter, 50 or 100 μA FSD (any meter
            of this sensitivity will do)
            Plastic box, approx. 150 by 80 by 50 mm
            Connecting wire, coloured

65 A vertical aerial for 70 cm

Introduction

If the range of your hand-held transceiver is very limited, what you need is a vertical aerial mounted outside. This design is a half-wave dipole fed at the end instead of the middle. Because the impedance of the dipole is high at its ends (and low in the centre), a matching circuit is needed so that this high impedance can be matched to that of the low-impedance coaxial cable. All that is needed is a coil, which increases the *electrical length* of the aerial to $\frac{5}{8}$-wavelength. (Note that the *electrical* length (i.e. the length as it appears to an RF signal) is not necessarily the same as the *actual* length.)

Construction

The aerial element and the coil are made from a single piece of 1.5 mm welding (brazing) rod, and the dimensions are given in **Figure 1**. Wind the coil around a 4 mm rod or the shank of a twist drill. The lower end is filed to a point and then soldered into the centre conductor of a 4-hole panel-mounting BNC socket. (Try to obtain a good-quality BNC socket with PTFE insulation – the insulation of cheaper sockets is easily damaged.) Trim the element to 427 mm (top of element to top of coil) *after* the wire has been soldered to the socket.

The base coil causes the aerial to be rather 'whippy', so a piece of 5 mm plastic knitting needle can be cut to the length of the coil and then forced into it.

The radials are made from four lengths of 3 mm welding rod. These are bent and soldered into the four mounting holes of the socket, and then cut to the lengths shown in Figure 1.

Testing

Using a standing-wave-ratio (SWR) meter connected between your aerial and the transceiver, measure the SWR (or obtain some indication of the reflected power) on transmit. If it is greater than 1.5, switch off the transmitter, trim about 3 mm off the end of the element, and try again. Repeat the process until the measured SWR (or the reflected power) is as low as you can get it.

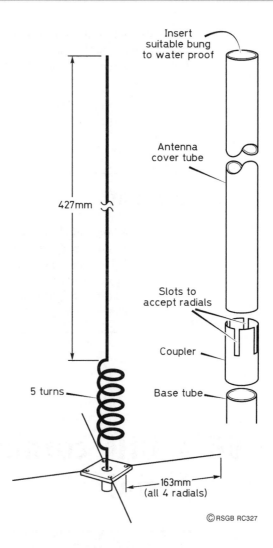

Figure 1 Construction of the 70 cm antenna

Installation

The aerial is made waterproof by enclosing it in a 22 mm diameter PVC waste water pipe. It is 'weldable', and available at plumbers' merchants, usually by the metre. A coupler is slotted to take the radials (Figure 1). File the BNC socket as required, so that it slides inside the coupler until the radials poke out of the slots. Cut a length of plastic tubing which is 30 mm longer than the aerial, and push it into the coupler. You will then need a plastic bung or screw-top to waterproof the top end.

Plastic welding solution is now applied to the joints, sealing the aerial inside the tube. The coupler has been weakened as a result of making the slots, so it is worthwhile applying PVC tape around this joint until the welding solution sets.

A support for the aerial is made from an off-cut of tube pushed into the lower end of the coupler and held with a self-tapping screw.

Parts list

1	4-hole panel-mounting BNC socket
1 m	3 mm brazing rod (may be available from a small garage)
1 m	1.5 mm brazing rod
1 m	22 mm PVC waste water pipe
1	22 mm straight coupler
	Plastic welding solution

---

66 A UHF corner reflector aerial

Introduction

The corner reflector is a well-known design and is capable of good performance on the VHF and UHF bands. At UHF, the practical implementation of the corner reflector is an ideal constructional project.

Some details

Expressed quite simply, the aerial consists of a $\frac{\lambda}{2}$ dipole (where λ is the standard symbol for wavelength, making a '$\frac{\lambda}{2}$ dipole' a half-wave dipole).

Nothing new in that, you might say. However, the interesting feature is the reflector, which is not the usual single element, but a 90° metal 'corner', acting rather like a parabolic dish as used for satellite signal reception. The wind resistance of this type of reflector makes it impractical so, to reduce the 'windage' quite significantly, we make the 'corner' from closely spaced rods, as illustrated in **Figure 1**.

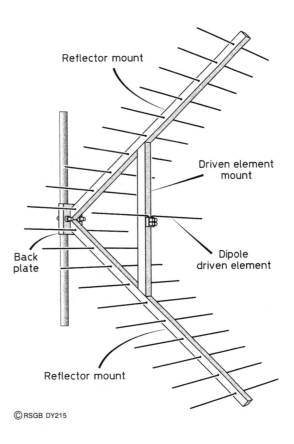

Reflector mount

Driven element mount

Dipole driven element

Back plate

Reflector mount

©RSGB DY215

Figure 1 70 cm corner reflector antenna

The reflector consists of a number of 0.6λ rods, spaced from each other by 0.1λ. The aerial frame can be made of metal or wood, but wood is easier to work with, and mounting the elements to the frame is simpler. The prototype was made with wood of 20 mm by 15 mm cross-section, as **Figure 2** shows. The wood was varnished for protection. The elements were made from 1.5 mm diameter copper wire, because a large reel of the wire happened to be available. The wire diameter is not critical; tubing could be used just as successfully. 14 SWG hard-drawn copper aerial wire would be even better than that used in the prototype.

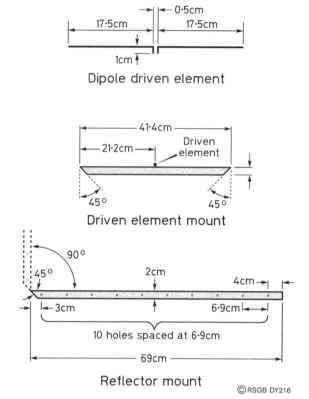

Figure 2 Driven element dimensions, together with boom dimensions for driven element and reflectors

Construction

This project is just as much a woodworking project as a radio project! Follow the instructions carefully, and you should have little trouble.

- Cut the booms for the reflectors, as shown in Figure 2. A mitre block is invaluable here in producing the 45° corners.
- Using the dimensions given on the diagram, mark the hole positions for the reflector elements, and then drill holes of a size which holds the elements firmly.
- Cut the driven element boom according to the diagram, and mark the point midway along the longer side, which will assist you later in positioning the driven element.
- Cut the back plate to size (about 120 mm by 80 mm). You may need to alter this size depending on the size of the U-bolt you will be using to clamp the aerial to the mast.

- If you want to be extra cautious in your construction, use the belt-and-braces approach, commonly known as 'screw-and-glue' to fix the booms to each other and to the back plate.
- Fix the reflector booms to the back plate first, then slide in the driven-element boom until it will go no further, then apply the wood glue and screw the two ends tightly to the reflector booms. Leave for the period prescribed by the glue manufacturers for the glue to harden.
- Varnish the whole structure.
- Cut the driven element to the correct size plus a couple of centimetres (the reason for this will be evident in the *Testing* section), and fix it to the centre of its boom (at the position you marked earlier) with a 'chocolate block' connector to which the coaxial feeder cable will eventually be connected.
- Cut and fix the reflector elements in place. If you find that these are a loose fit in the holes then, *for each element*, drill a pilot hole through the boom to intersect the hole for the element. File off the point of a woodscrew, and screw it gently into the pilot hole until it meets the element and grips it in place. You will now see why the point was filed off! Alternatively, you can glue the elements in place.

Testing

Place the aerial on a mast, clear of obstructions. Connect it to a transceiver with a length of coaxial cable, with a standing-wave-ratio (SWR) meter in circuit. Find a clear frequency, identify your transmission and ask if the frequency *really is* clear. If so, key the transmitter again and note the SWR. **Do not stand in front of any aerial when it is radiating!** The length of the driven element must be adjusted to obtain an SWR of less than 2. If you have to shorten the dipole, bend the ends over rather than cut them off. That way, if you go too far, your can lengthen them again! The dipole was initially cut too long intentionally, to allow for adjustment here! Bending the ends over also reduces the risk of physical damage to clothing, skin and eyes. You may like to consider applying the same technique to the reflector elements for that reason alone.

Moving on . . .

Once you have warmed to the idea of the corner reflector as an aerial, you might like to ring the changes regarding the reflector. How about replacing the 20 reflector elements with a wire mesh, such as garden centres sell as 'chicken wire'? Choose the finest mesh if there is a choice. Some extra support may be needed around the edges of the mesh, but you could go on to make a comparison of aerial gain between the two types, using the *UHF Field Strength Meter* described elsewhere in this book.

Materials

Stiff wire or thin-walled tubing for dipole and reflector
Frame – wood, 15 mm by 20 mm cross-section, lengths given in text
Back plate – stout plywood, dimensions given in text
U-bolt to suit mast
Wood screws
Wood glue
50 W coaxial cable for feeder
2-terminal 'chocolate block' for dipole connection to feeder
Varnish

67 A switched dummy load

Introduction

A dummy load is a pure resistor of value 50 ohms which can replace your transmitting aerial and enable you to operate the transmitter for test purposes without radiating a signal. It sounds simple enough, but there are two main problems. Firstly, it is impossible to delve into your junk box and emerge with a resistor that will dissipate 100 W PEP and still retain its marked resistance value. Secondly, a 'pure resistance' is very difficult to achieve. A pure resistance is a device which has resistance but no reactance. All common resistors have significant reactance at radio frequencies, particularly the wire-wound varieties, which have a helical (i.e. wound like a coil) construction. This is particularly annoying, because wire-wound construction is normally used for large-wattage resistors.

Although all resistors have *some* reactance, not all are quite as bad as the wire-wound type. Carbon film resistors are made by depositing a thin film of carbon on the surface of a small, hollow ceramic cylinder, the thickness of the film of carbon determining the value of the resistor. Provided the lead lengths are kept short, these resistors have a tolerably small reactance, and will be used in this project.

Bearing the load

A 2 W carbon film resistor is hardly going to withstand our 100 W PEP of SSB, so it is obvious that the design of our dummy load must be a little more complex than a single resistor and a switch, despite what **Figure 1** might suggest! In fact, it uses 20 resistors, each of value 1000 Ω (1 kΩ). How does this solve our problem?

Perhaps a little theory is in order here, but no more than is required by the Radio Amateurs' Examination.

When two *equal* resistors of value r are combined in parallel (i.e. side by side), the total resistance, R_T, is given by:

$$\frac{1}{R_T} = \frac{1}{r} + \frac{1}{r}.$$

Adding $1/r$ to $1/r$ gives $2/r$, therefore:

$$\frac{1}{R_T} = \frac{2}{r}, \quad \text{i.e. } R_T = r/2.$$

So, by connecting *two* equal resistors in parallel, we get a combined resistance which is *half* the individual resistances. If we combine *three* in parallel, we get a *third* of the resistance, and so on.

Here, we are connecting 20 resistors of 1 kΩ in parallel, so we will produce an overall resistance of *one-twentieth* of the individual resistance, i.e. 1000/20 = 50 Ω, which is what we set out to achieve!

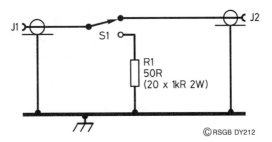

Figure 1 Basic circuit diagram

This is not the only advantage, though. Each resistor is capable of dissipating 2 W, so 20 of them will safely dissipate 40 W, for short periods at least. This power dissipation is approximately the same as 100 W PEP of normal speech, so the design should be capable of use in an 'average' amateur station. The power-handling ability of any dummy load can be improved by providing a 'heat sink' which helps to conduct the heat energy away from the resistors, thus lowering their temperature. One popular heat sink is a can of transformer oil, into which the resistors are immersed. This design uses a rather more mundane heat sink, but which is adequate for the job in hand.

Figure 2 PCB resistor
supports

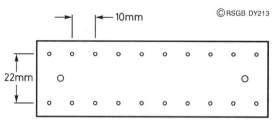

©RSGB DY213

10mm

22mm

2 PCBs (single sided) 100 x 35mm, 20 holes
1 'earth' PCB with two 3mm holes

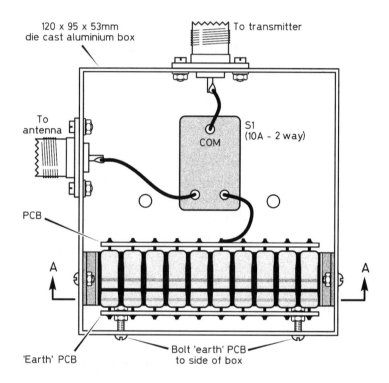

120 x 95 x 53mm
die cast aluminium box

To transmitter

To
antenna

S1
(10A - 2 way)

COM

PCB

A

A

'Earth' PCB

Bolt 'earth' PCB
to side of box

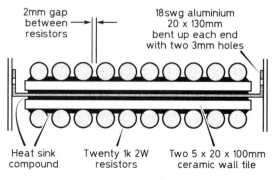

2mm gap
between
resistors

18swg aluminium
20 x 130mm
bent up each end
with two 3mm holes

Heat sink
compound

Twenty 1k 2W
resistors

Two 5 x 20 x 100mm
ceramic wall tile

Figure 3 Switched dummy
load, component layout

Section A-A

©RSGB DY214

Pouring the heat away

Two basic PCBs are needed, the details being given in **Figure 2**. Both measure 11×35 mm. Each has a set of 20 holes of diameter about 1 mm to take the ends of the resistors. Only one of them has two 3 mm holes which are used for mounting the completed load. **Figure 3** shows how the heat sink is assembled. Solder one end of each resistor into the 'earth' PCB, and then mount this to the box using the two bolts as shown in the diagram. Between each rank of resistors is a 'sandwich' consisting of an aluminium strip and two pieces of ceramic wall tile, to act as a heat sink for the resistors. Heat sink compound is used to provide good thermal contact between the resistors and the tiles, and between the tiles and the aluminium strip. This is shown in the lower part of Figure 3. Thread the loose wires of the resistors through the holes in the second PCB, solder into place, and crop the protruding wires.

Switching

A changeover switch must be used so that the transceiver can be switched between the dummy load and the aerial without unscrewing connectors. For most purposes, an ordinary 10 A 230 V changeover switch will suffice, as found in many electrical shops and DIY stores.

Wire this into the circuit as shown in Figure 1 and Figure 3. Check your wiring. Put S1 in the 'dummy load' position. If you have a multimeter which includes an ohmmeter, measure the resistance across the socket, J1, before connecting it to the transceiver. It should be very near 50 Ω. With S1 in the 'aerial' position, there should be an infinite resistance across J1. Move your ohmmeter to read the resistance across J2. It should be infinite for both positions of S1. If all seems correct, close the box and your dummy load is ready for use!

Parts list

Resistors
    1000 Ω (1 kΩ) 2 W carbon film, 20 required
Additional items
    10 A 230 V changeover (SPDT) switch
    SO–239 sockets, 2 required
    Aluminium box $120 \times 95 \times 53$ mm or similar
    Ceramic wall tile cut as required
    Aluminium strip, 18 SWG
    Nuts and bolts as required

68 A simple Morse oscillator

Introduction

This is an excellent project which uses the 'junk box' as its source of components. If you have trouble in finding the bits for this one, a good source of the components for this and many other similar projects is to be found with the parts list at the end of the project.

The circuit

This is shown in **Figure 1**, and uses an 'unknown' Plessey chip, which makes the overall circuit extremely easy to build. A 0.1 μF capacitor is connected between pins 7 and 8, a speaker (in the popular 8 to 25 Ω impedance range) is connected between pins 8 and 9. If a 9 V battery is connected with its positive terminal to pin 8 and its negative terminal to pin 1, 3 or 5, a tone will be produced in the speaker.

To make this circuit into a good Morse practice oscillator, it is necessary only to insert a Morse key into the supply rail from the battery.

However, there is another refinement which you may care to build into the circuit. The tone from the loudspeaker is different, depending upon which of pins 1, 3 or 5 you use. In the prototype, a single-pole changeover switch was used to select the tones from pin 1 or pin 3, and the Morse key would be connected to the circuit via a small jack socket. There is no need for an on/off switch, as the Morse key performs that function. The switch and the jack socket can be seen in the photograph.

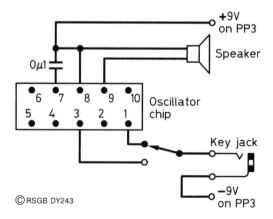

Figure 1 Morse oscillator, circuit diagram

© RSGB DY243

238

The case

Almost any small loudspeaker will do; the higher the impedance the better. The 'impedance' is the figure usually printed on the rear of the speaker magnet. It is a value given in ohms (Ω). For simple circuits like this, you will usually find that the higher the impedance (within reason), the louder the sound it will produce. Speakers from old transistor radios will work, although their impedances can be rather low sometimes.

Any case big enough to house the components and the battery will be suitable. The prototype used a 'Walkman'-type speaker and case, and is shown in the photograph.

To use the circuit, simply attach a 3.5 mm jack plug to your Morse key, and insert it into the socket. Nothing should happen until you press the key, when a tone should be heard from the loudspeaker.

Another application

Try soldering the two wires of a 'twisted pair' to the 3.5 mm jack plug. Touching the wires together produces a tone from the speaker. This simple circuit can then be used as a 'continuity tester'. Touch the two wires to the ends of a fuse. If there is no sound, then the fuse is blown. There are many other such tests you could perform with this device – to check whether there really *is* a connection between one end of a wire and the other, for example.

Warning

You must **never** make such tests on any equipment which is connected to the mains supply, even if it is switched off. If you want to make any such tests, make sure you are supervised by someone who understands what you are doing and is competent to advise and supervise you.

Parts list

Plessey oscillator chip
0.1 μF capacitor (disc ceramic)
Loudspeaker 8 to 25 Ω
3.5 mm jack socket (and 3.5 mm jack plug if needed)
Single-pole changeover switch

The Plessey oscillator chip is available from J. Birkett, 25 The Strait, Lincoln LN2 1JF, tel: 01522 520 767.

John Birkett may also provide a kit of parts (the chip, capacitor and loudspeaker).

---

69 A bipolar transistor tester

Introduction

This is a circuit which will test normal transistors, i.e. npn or pnp. It has the advantage of being able to test devices while they are still connected in their original circuits. However, when such tests are made, **the circuit containing the transistor under test must not be switched on**.

The circuit and how it works

The circuit runs from a 9 V battery such as a PP3 or six AA-type 1.5 V cells. Alkaline cells are to be preferred, as their electrolyte leakage properties are better. The circuit shown in **Figure 1** uses a single CMOS integrated circuit type 4001 or 4011. CMOS circuits require special handling precautions which are described in the project *Christmas Tree LEDs*, elsewhere in this book.

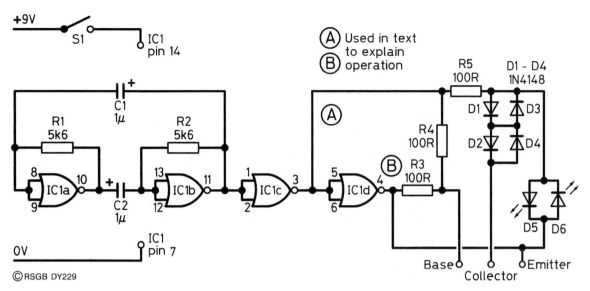

Figure 1 Transistor tester, circuit diagram

Inside the IC are four logic gates (see *Digital Logic Circuits*) which are all connected as inverters, which means that the output signal is always the logical 'opposite' of the input. The first two gates are connected as an oscillator; the circuit being the same as that used in *An Electronic Die*.

The output of the oscillator, at pin 11, is connected to the input of a buffer stage, IC1c, which helps to isolate the oscillator from the circuit that follows it. The buffer output appears on pin 3, which we shall label as *test point A* for future use. Another inverter, IC1d, follows this, its output at pin 4 being labelled *test point B*.

There are two LEDs connected back to back at the circuit output. These are D5 and D6, D5 being red and D6 being green. Whatever the output of the oscillator at any instant, one of the LEDs must be lit and the other unlit. With point A positive and point B zero, the red LED is lit, when A is zero and B is positive, the green LED is lit. Because the oscillator output is repeatedly switching from one polarity to the other, the lit LED is alternately red and green. They switch between the two colours much faster than we can see, so what we think we see are both LEDs lit together.

The two 100 ohm resistors, R3 and R4, provide the bias to the base of the transistor under test. When A is positive and B is zero, the base-emitter junction of the transistor will be forward-biased, and the transistor will switch on (if it is a working npn type). When the transistor is on, it effectively short-circuits D5 (the red LED) and it extinguishes. When point A is zero and B is positive, an npn transistor will be switched off and the green LED (D6) will light. Thus, for a working npn transistor, only the green LED is lit.

If the transistor under test is a pnp type, it will switch on when A is zero and B positive, thus short-circuiting the red LED (D6). When A is positive and B zero, the transistor is off and the red LED (D5) is lit. Thus for a working pnp transistor, only the red LED is lit.

To summarise, the states of the LEDs indicate the following conditions:

- Both LEDs apparently lit: no transistor connected, or transistor permanently open circuit.
- Neither LED lit: a collector-emitter short-circuit is almost certain.
- Red LED alone: pnp transistor in working order.
- Green LED alone: npn transistor in working order.

Construction

The prototype tester was built on a piece of Veroboard measuring 15 strips by 18 holes. Cut the tracks using a track-cutting tool or a 3 mm ($\frac{1}{8}$ inch) twist drill, as shown in **Figure 2**. Notice that, in this diagram, there is no

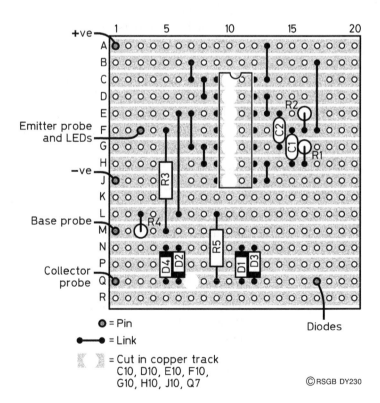

Figure 2 Transistor tester, component layout

242

track 'I', so try to avoid miscounting when you translate diagram positions to real positions on the board. Hold the board up to a strong light to ensure that there is no copper swarf shorting adjacent tracks together, and that you have made the cuts in the correct places.

Having done this, insert and solder Veropins for all the connections to be made to components not on the board itself. Then solder in the IC holder and the wire links. Then insert the other components in the order resistors, capacitors and diodes D1–D4. Some resistors are mounted vertically so that their connections are on adjacent tracks. Double check the diode polarities – it is easy to make a mistake when wiring diodes in anti-parallel! Next, connect up the off–board components, again making sure that the LEDs have the correct polarity. The probe leads for the emitter, base and collector should be made from different colours of wire and terminated in probe clips (small insulated crocodile clips).

Check carefully for dry joints and errant blobs of solder. Plug the IC into its holder, ensuring that it is inserted the right way round, as shown in Figure 2.

Testing

Without a transistor in circuit, and the battery connected, both LEDs should be lit. Connect a known good npn transistor and verify that the green LED lights. Now simulate two transistor faults: disconnect the base lead and both LEDs should light; remove the transistor and connect the emitter and collector leads together. Neither LED should light.

Repeat the tests with a known good pnp transistor. The results should be the same, except that the correct indication should now be a lit red LED. On your computer, make a small label of the bulleted list above, to fit on your tester showing the states of the LEDs and what they mean. It will act as a useful *aide mémoire* when you use the tester in future.

Using

The circuit will test transistors in isolation or in an existing circuit, i.e. prior to use. You can check the lead identifications in component catalogues such as the Maplin catalogue. The tester is ideal for going through the large bags of unmarked transistors that you can buy for a song at rallies. You can sort them into three piles – npn, pnp and dud!

Parts list

Resistors: all 0.25 watt, 10% tolerance or better
    R1, R2   5600 ohms (Ω)
    R3, R4, R5   100 ohms (Ω)

Capacitors
    C1, C2   1 microfarad (μF) electrolytic, 16 V WKG

Semiconductors
    D1–D4   1N4148 general-purpose silicon diodes
    D5   Red LED
    D6   Green LED

Integrated circuit
    IC1   CMOS 4001 or 4011

Additional items
    Veroboard, 15 strips by 18 holes
    Veropins
    PP3 battery and connector (or 6 × AA cells in PP3 clip holder)
    Switch, SPST
    Connecting wire
    One each of known working npn and pnp transistors for test
      purposes.

Source

Components are available from Maplin.

70 The 'Yearling' 20 m receiver

Introduction

Published to celebrate the first anniversary of *D-i-Y Radio*, this excellent receiver design forms a suitable 'second receiver project' for those who have successfully completed the MW receivers earlier in this series. The receiver is powered from a PP3 battery or from a mains adaptor, and can be built with the help of an experienced constructor, on a prototype board. The circuit diagram and some of the components used are shown in the separate diagram. Headphones or a loudspeaker can be used and, once the radio is completed, a few simple adjustments will make the *Yearling* spring to life!

Building the receiver

Before starting the constructional process, start by identifying all the parts. One by one, tick them off against the parts list. Are their values correct? The varactor diode is a twin type (see circuit diagram overleaf), and must be cut *carefully* down the middle, producing two devices, D1 and D2, with two wires each.

First, solder the IC sockets, followed by the coils (inductors); L1 is pink inside the top, and L2 is yellow inside. Then, solder in the varactors; the lettering on D1 should be next to coil L1, and the lettering on D2 should face resistor R7. After those, the capacitors, wire links and resistors should be soldered to the board. Take care to wire the voltage regulator, IC3, correctly. Solder in the crystal X1 as quickly and deftly as you can – crystals do *not* take kindly to having their leads bent and being fried with a soldering iron! Make sure that the electrolytic capacitors C2, C12, C15 and C16 are fitted the right way round. Most electrolytic capacitors have only the negative lead marked.

Figure 1 shows the rear of the front panel, illustrating the connections from the board and antenna socket to the controls. All normal connecting wires are 22 SWG or thereabouts, with insulation. Their lengths should be about 15 cm, except for the battery lead to the switch, which is about 8 cm. It is recommended that you use different-coloured wire for each connection to a control. **Figure 2** shows this. The variable resistor section of VR5, the AF gain control, uses single screened cable connected to 0 V (ground) at the

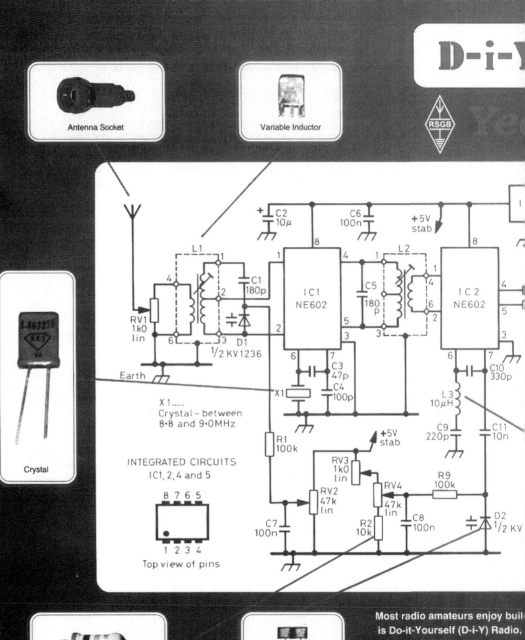

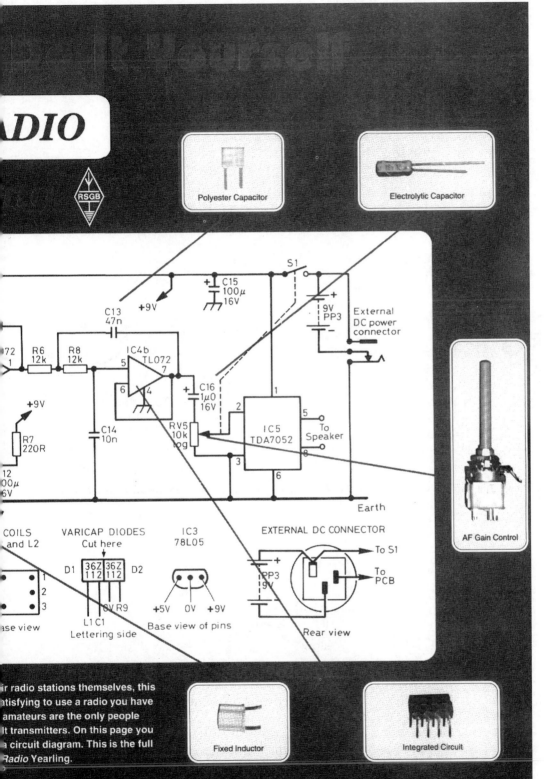

Polyester Capacitor

Electrolytic Capacitor

RSGB

+9V

C15
100µ
16V

S1

+
9V
PP3
−

External
DC power
connector

C13
47n

972
1

R6
12k

R8
12k

IC4b
TL072

5

7

6

4

C16
1µ0
16V

2

1

5

To
Speaker

IC5
TDA7052

8

RV5
10k
log

3

6

+9V

R7
220R

C14
10n

12
00µ
6V

Earth

AF Gain Control

COILS
and L2

VARICAP DIODES
Cut here

IC3
78L05

EXTERNAL DC CONNECTOR

To S1

D1

36Z
112

36Z
112

D2

PP3
9V

+

To
PCB

1

2

3

V R9

L1 C1
Lettering side

+5V   0V   +9V

Base view of pins

PP3
9V

+

ase view

Rear view

r radio stations themselves, this
atisfying to use a radio you have
amateurs are the only people
t transmitters. On this page you
a circuit diagram. This is the full
Radio Yearling.

Fixed Inductor

Integrated Circuit

e Road, Potters Bar, Herts. EN6 3JE

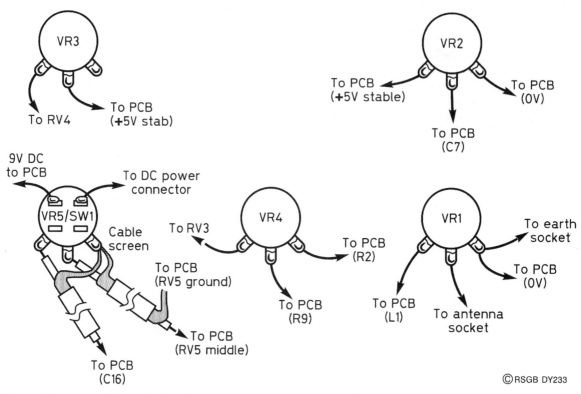

Figure 1 Rear view of the variable resistors. Check the connections carefully to make sure the wires fit the correct holes

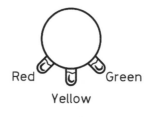

©RSGB DY234

Figure 2 It is helpful to use different colour insulated connecting wires. Wires between each variable resistor and the board should be twisted together to give a neat wiring outfit.

board end. Now fit the ICs into their sockets. Make sure that they are the right way round (see circuit diagram) and that each pin lies directly above its corresponding socket before applying *gentle* pressure with the back of the board firmly supported.

Check that all the connections are correct (don't *assume* this – check the ends of each wire against the circuit diagram) and that all your soldered joints are shiny. Lastly, drill the five 10.5 mm diameter holes for the main controls. On the side of the case, a 6.3 mm diameter hole is needed for the speaker socket, and 8 mm holes for the antenna and earth sockets. The external power socket requires an 11 mm diameter hole.

Adjusting it

Before you can make adjustments, and in order to hear *anything* on your *Yearling*, you will need to connect an antenna (aerial) to the antenna socket. About 8 metres of wire, preferably outdoors and as high as possible, is all you need to connect to the socket. Connect the 3.5 mm jack socket for the speaker, and a 3.5 mm jack to your speaker leads. Connect a battery and switch on.

1. Using a very small screwdriver or, better still, a non-metallic 'trimming tool', *gently* screw in the core of L2 as far as it will go, but *don't force it*. Then unscrew it by three turns anticlockwise.
2. Set VR1, VR2 and VR4 to mid-position and rotate the core of L1 anticlockwise until the hissing noise you hear reaches a maximum intensity. Then adjust L2 for maximum noise.
3. If you now tune carefully with the main tuning control, VR4, you should hear some amateur Single-Sideband (SSB) speech signals. You may have to adjust the bandspread (fine tuning) control, VR3, to make the speech sound normal.
4. Having verified that everything is working, switch off and mount the controls on the front panel and the sockets on the side. To do this, it is much safer to disconnect all the controls and sockets, mount them in their final positions, and then wire them up again.
5. Fit the front panel knobs, connect your aerial and switch on again, checking that everything is working. Then, locate the cluster of CW (Morse) signals to be found at the bottom of the 20 metre band, set the bandspread control to mid-position and slacken off the main tuning knob. Turn the knob (but *not* the control!) until the pointer lies a little clockwise of 14.0 MHz. The SSB signals should now lie roughly between the dial centre and 14.35 MHz. Tighten up the knob.
6. Finally, fix the board to the rear panel, and secure the battery (if you are using one). Attach the rear panel to the back of the box, and you are finished!

Listening!

Make a habit of keeping the bandspread control in its centre position when searching for stations; then you can adjust it either way to make the signals readable. Unless you have a very big aerial, it is best to have the 'RF Gain' control, VR1, at maximum. Use the 'Antenna Tune' control, VR2, to give the best signal, and control the volume with the 'OFF/AF Gain' control. You will find some excellent DX stations with your *Yearling* receiver, and it will serve you well.

Parts list

Resistors (all 0.25 W, 5%)
R1, R5, R9	100 kilohms (kΩ)
R2	10 kilohms (kΩ)
R3, R4	1.5 kilohms (kΩ)
R6, R8	12 kilohms (kΩ)
R7	200 ohms (Ω)

Capacitors (all rated at 16 V or more, tolerance *at least* what is quoted)
C1, C5	180 picofarads (pF) polystyrene 5%
C2	10 microfarads (μF) electrolytic
C3	47 picofarads (pF) polystyrene 5%
C4	100 picofarads (pF) polystyrene 5%
C6, C7, C8	100 nanofarads (nF) or 0.1 microfarad (μF) ceramic
C9	220 picofarads (pF) polystyrene 2%
C10	330 picofarads (pF) polystyrene 2%
C11, C14	10 nanofarads (nF) or 0.01 microfarad (μF) ceramic
C12, C15	100 microfarads (μF) electrolytic
C13	47 nanofarads (nF) or 0.047 microfarad (μF) polyester, 5%
C16	1 microfarad (μF) electrolytic

Variable resistors
VR1, VR3	1 kilohm (kΩ) linear
VR2, VR4	47 kilohms (kΩ) linear
VR5	10 kilohms (kΩ) log with switch

Inductors
L1	Toko KANK3335R
L2	Toko KANK3334R
L3	10 microhenries (μH), 5%

Semiconductors
IC1, IC2	Philips/Signetics NE602 or NE602A
IC3	78L05 5 V 100 mA regulator
IC4	TL072 Dual Op-Amp
IC5	Philips TDA7052 audio amplifier

Additional items
D1, D2	Varactor diode Toko KV1236 (cut into two sections – see text)
X1	Crystal 8.86 MHz type (from Maplin, etc.)
4 off	8-pin DIL sockets for IC1, IC2, IC4 and IC5
2 off	4 mm sockets aerial (red) and earth (black)
1 off	3.5 mm chassis-mounting speaker jack socket
1 off	DC power socket for external supply (if required)
4 off	Red knobs with pointers
1 off	Tuning knob with pointer (e.g. 37 mm PK3 type)
1 off	Printed-circuit board or prototype board
1 off	Plastic case approx 170 × 110 × 6 mm (e.g. Tandy number 270-224)
1 off	Speaker 8–32 Ω impedance (or headphones)

71 Adding the 80 metre band to the Yearling receiver

Background

You will have noticed that your *Yearling* receiver has a dial which shows coverage of the 80-metre amateur band (3.5–3.8 MHz). This band is used for local contacts during the day, and contacts up to about 1600 miles in darkness. Longer distances are possible, particularly in the middle of winter.

The modifications

Only a few extra parts are required, as you may have noticed from the parts list. A low-pass filter, FL1 (one which passes low frequencies and rejects higher frequencies), is switched into the circuit on 80 m. The circuit of the

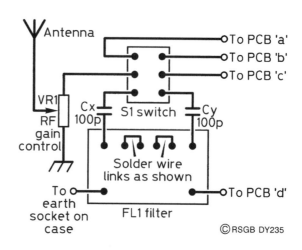

Figure 1 The circuit diagram shows the extra components for 80 m operation. Note the connections to the PCB 'a', 'b', 'c' and 'd'

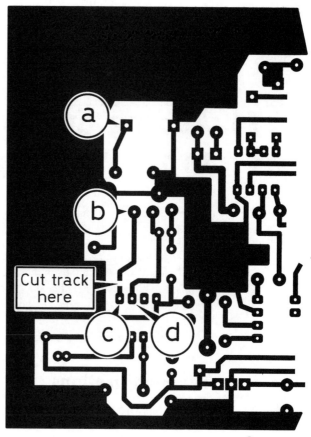

Figure 2 The underside of the PCB. Wires are connected from the switch and filter as shown

switch and its connections is shown in **Figure 1**. Before making the modifications, dismantle the receiver so that you have easy (and safe) access to the case and the track side of the PCB.

1. The first thing to do is to drill a 6.5 mm diameter hole in the side of the case into which the switch fits.
2. Then, using a sharp Stanley knife or scalpel, carefully cut the track on the PCB as shown in **Figure 2**, making a gap about 1 mm wide.
3. Using 10 cm lengths of different-coloured insulated wire, make the four connections, a, b, c and d, to the PCB, as shown in Figure 2.
4. Solder the two links on the filter, and then make the connections to the switch, capacitors and PCB. You will have to disconnect the existing wire between the RF gain control, VR1, and pin 4 of L1 on the PCB.
5. Lastly, check your new connections carefully, then mount the switch in the new hole and fix the filter to the bottom of the case with a little glue, as shown in **Figure 3**. Reassemble the circuit, and replace the back of the case.

More testing!

Firstly, switch your new switch, S1, into the 20 m position, to check that the original circuit still works! If you find that the 'Antenna Tune' control peaks at a slightly different position, don't worry.

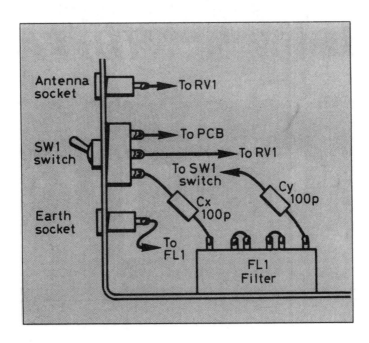

Figure 3 Internal view of the Yearling case. The filter FL1 is attached to the base with glue

Now switch to 80 m, and tune around the anticlockwise end of the dial; you should hear some SSB stations, particularly in the evenings and at weekends, when may people are on the air. At the other end of the travel of the tuning control, you should hear CW (Morse) stations.

The Radio Society of Great Britain broadcasts amateur radio news every Sunday morning on or about 3.65 MHz; the table below has the details. Finally, a good antenna is more important than ever for 80 metre reception – aim for more height and length, and then consider the project concerned with making an Antenna Tuning Unit (ATU)!

Parts list

Capacitors (all rated at 16 V or more, tolerance 10% or better)
    Cx, Cy   100 picofarads (pF) polystyrene

Filter
    FL1   Toko 237LVS1110 low-pass filter

Additional items
    SW1   2-pole 2-way (changeover) toggle switch
    7 off   Short lengths of insulated wire of different colours

---

The RSGB news broadcasts, GB2RS – Sunday mornings

Frequency (MHz)	Local time	Reception area
3.650	0900	SE England
3.650	0930	Midlands
3.650	1000	SW England
3.650	1100	Yorkshire
3.640	1130	Aberdeen
3.660	1130	Glasgow

The Midlands transmission is repeated at 1800 (6pm) local time on 3.650 MHz. All frequencies are approximate in order to avoid interference, and use lower sideband (LSB). If you also have a 40 metre receiver, there are GB2RS news broadcasts on 7.048 MHz at 0900 local time from Northern Ireland and from 1100 local time from the north midlands.

72 How the Yearling works

Introduction

The *Yearling* was designed to provide an introduction to *Amateur Radio* on the 20 m amateur band. Let's look at how the different sections (or 'stages') of a radio work, and how they fit together to form a complete receiver. **Figure 1** shows a block diagram which you can follow and compare with the circuit diagram of your *Yearling* receiver.

The antenna (or aerial)

Connected to your receiver, it will pick up not only amateur signals, but all other signals as well! This means that the receiver has to select the one signal that interests you, while rejecting all the others. The following stages do just that.

The RF filter

This stage (centred around L1) selects the band of radio frequencies (RF) containing the signal you want, in this case, those having wavelengths around 20 m. Signals from the 40 m band, for example, would not get through.

The crystal oscillator

This is an oscillator circuit designed around a quartz crystal (X1), and has a very stable frequency. It produces a single, very pure frequency to feed into the mixer. A crystal having a frequency between 8.800 MHz and 9.000 MHz is suitable for this circuit. The oscillator and mixer functions are both carried out inside IC1.

The first mixer

Yes, this stage 'mixes' two signals together. In this case, the two signals are (i) from the aerial via the RF Filter, and (ii) from the crystal oscillator. Two

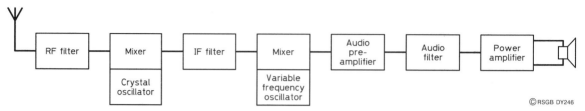

Figure 1 Block diagram of the Yearling, showing how the various stages fit together to make a complete radio receiver

bands of signals emerge from the mixer. The first is centred upon a frequency equal to the incoming signal frequency **added** to the crystal frequency, and the second is centred upon a frequency equal to the incoming signal frequency **subtracted** from the crystal frequency. Look at an example – if the signal is at 14 MHz and the oscillator at 9 MHz, then the mixer outputs will be 14 + 9 = 23 MHz and 14 – 9 = 5 MHz.

Intermediate frequency (IF) filter

It is the purpose of the IF filter (centred around L2) to select only one of these two bands of frequencies emerging from the mixer. In this case, it is the lower band of frequencies (around 5 MHz) which we select. This is because, in general, lower frequencies are easier to handle than higher ones.

Variable-frequency oscillator (VFO)

The VFO (part of IC2) enables us to tune into a particular station, and operates over a band of frequencies between 5 MHz and 5.35 MHz in this receiver. You will notice that IC1 and IC2 are the same type of chip, so that you will be expecting another mixer stage to be associated with the VFO. You are quite right!

The second mixer

This mixer obeys exactly the same rules as those of mixer 1. Sum and difference frequencies are produced, like this. Mixing is between the incoming IF signals (around 5 MHz) and the VFO signals (around 5 MHz), producing output frequency bands centred upon 10 MHz and 0 MHz. The use of the words 'band of frequencies' throughout this explanation is intentional. If all the signals were pure, there would be no bands; the bands are produced because of one thing – the modulation imposed on the pure frequencies at the transmitter. So, the 'bands' contain the one thing that we want to extract from the signal, and that is the speech or Morse code that the signal contains. The band of frequencies at 0 MHz is just that – the audio frequencies we want in the loudspeaker. Because of this, the audio frequency output of the second mixer is selected and passed on for amplification.

The audio preamplifier

Preliminary amplification of the minute audio signal which emerges from the second mixer is provided by IC4a, which will respond only to audio signals, automatically rejecting the 10 MHz signal.

The audio filter

The bandwidth of normal speech when transmitted by an amateur station is around 3 kHz, so there is no advantage to be gained in amplifying frequencies greater than this. IC4b is known as a low-pass filter, because it passes (lets through) lower frequencies and rejects higher ones.

The power amplifier

IC5 produces the final audio amplification and provides enough power (about 350 milliwatts (mW) to drive a small speaker.

How does it work on 80 m?

If you have fitted the 80 m modification to your receiver, you are probably wondering how the circuit works at this different frequency. Firstly, the filter which you fitted selects the 80 m band instead of the 20 m band. The only other slight difference lies in the way the first mixer stage works. Its job is to produce sum and difference frequencies from the incoming signal and crystal frequencies. On 20 m, it did this by subtracting the crystal frequency (9 MHz) from the incoming frequency (14 MHz) to produce an IF output of 5 MHz. On 80 m, the incoming frequency (3.5 MHz) is subtracted from the crystal frequency (9 MHz) to produce an IF output of 5.5 MHz, which is still within the tuning range of the VFO in the next stage.

73 A field strength meter

How it works

The basic field strength meter uses the circuit of a crystal set, but with a meter replacing the headphones. A better design, which is used here, is that of a voltage doubler, giving more sensitivity. **Figure 1** illustrates the voltage doubler circuit.

The AC input shown will be our RF input, which will be explained soon. Diode D1 will pass the positive half of the signal and use it to charge up C1 to the peak value of the signal. D2 uses the negative half of the signal to charge up C2 to the same value. Because C1 and C2 are in series, the peak voltage appearing across *both* of them (which is the DC output voltage) is equal to *twice* the peak input voltage, hence the name voltage doubler. If you're wondering why DC is present at the output when AC comes in at the

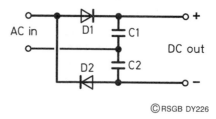

Figure 1 Voltage doubler circuit

©RSGB DY226

input, remember that capacitors pass AC (RF) but block DC; thus the RF is shorted out through C1 and C2, but the steady voltage (DC) remains across the two capacitors. The voltage is thus proportional to the size of the RF signal applied at the input.

The circuit

The voltage doubler is converted into a field strength meter using the circuit of **Figure 2**. A piece of wire serves as the aerial to provide an RF signal across the radio-frequency choke (RFC). A choke is an inductor which is large enough to prevent the RF passing through it – it 'chokes' the RF. This produces the maximum RF signal at the input to the voltage doubler, and the DC output from it is measured on the meter.

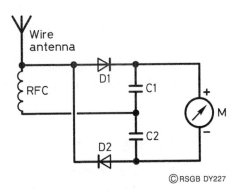

Figure 2 Field strength meter

The parts used are not very critical. The RF choke can have any value between about 1 mH and 2.5 mH. Almost any common diode such as the 1N914 or 1N4148 can be used for D1 and D2. The two capacitors could be any value between 1 nF and 100 nF (0.001 μF and 0.1 μF). The meter should be reasonably sensitive, with a full-scale deflection (FSD) in the range 50 μA to 100 μA. Look for VU meters at rallies – these are ideal.

Construction

The prototype circuit was made on matrix board. If you don't want to use pins with the board, simply push the component leads through the holes and make the connections on the underside of the board, either with the excess component leads themselves, or with ordinary connecting wire.

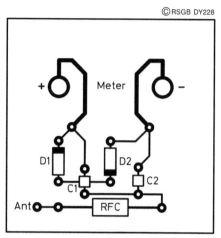

©RSGB DY228

Figure 3 Board layout and
interconnections

Component side of board shown.
All connections are under board

Figure 3 shows how the parts are placed and connected. The matrix board
can be mounted directly on the meter using its terminal bolts!

In use

Needing no power supply other than an RF signal, just connect it all up
and leave it to work! A short length of insulated wire is enough to pick up
some RF and display it on the meter. Using a 200 μA meter, about 3
metres of wire gave a good deflection on the meter. To increase deflection,
the wire can be wrapped around the aerial lead, provided that the wire is
PVC covered and doesn't come into contact with the aerial wire.

This requires a little experimentation. Try a long piece of wire first and
adjust its position until the meter needle kicks whenever there is a
transmission. It is very reassuring to see the meter moving during a
transmission. Although SWR meters also indicate power, they are usually
set to read reverse power, and show little or no movement during
transmission.

Parts list

D1, D2	1N914, 1N4148 or similar
C1, C2	10 nF disc ceramic
RFC	Miniature axial choke (1 mH)
Meter	Surplus VU meter or similar
Matrix board or similar	

Components are available from Maplin.

74 Preselector for a short-wave receiver

Introduction

A preselector is a simple RF tuned amplifier which is inserted between the aerial and the receiver. It provides some extra gain and may improve the overall performance of the receiver. This project uses a Field-Effect Transistor (FET) amplifier in grounded-gate mode.

The design has a tuned circuit at both the input and output which, with excessive gain and poor construction, would produce only one thing – oscillation! So, to avoid this happening, we will have only a low gain, and use a circuit which provides good isolation between input and output. The grounded-gate FET amplifier fulfils both these criteria. It will also cover a frequency range from about 7 to 30 MHz, which includes most of the HF amateur bands.

The circuit

This is shown in **Figure 1**. The signal from the aerial arrives at an RF transformer, the secondary of which is tuned with capacitor VC1a. The output from the tuned circuit is taken from a tap on the secondary to the source of the FET. The gate is grounded (earthed) and the amplified signal appears at the FET drain, which is then fed to the primary of another RF transformer, which is tuned by VC1b. The output to the receiver comes from the secondary of the RF transformer.

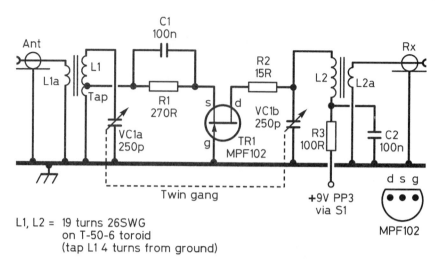

L1, L2 = 19 turns 26SWG
on T-50-6 toroid
(tap L1 4 turns from ground)

L1a, L2a = 3 turns

Figure 1 HF preselector, circuit diagram

©RSGB DY256

Notice that the two RF transformers are identical, but they are used 'back to back', with the secondary of the first and the primary of the second being tuned. They are tuned with identical capacitors, fitted on the same shaft of a variable capacitor. We say that the two capacitors are 'ganged'. Because L1 and L2 are the same, and VC1a and VC1b are the same, both RF transformers should be resonant at the same frequency, no matter what that frequency is.

Construction

The final layout should look something like that shown in **Figure 2**. The external connectors and controls being two SO-239 sockets for connection to your receiver and aerial, a tuning control and its associated scale, and an on/off switch.

The circuit can be put together on a plain matrix board, using pins to anchor the components, or simply by pushing the component leads through the board and making connections on the underside. The layout of the prototype is shown in **Figure 3**. Mounting the board to the aluminium box is accomplished with bolts, solder tags and stand-off insulators.

Check your construction against the circuit diagram and against the layout diagram. Wire in the PP3 battery clip, put the switch in the 'off' position, and fit the battery. Testing can be carried out without fitting the top of the box.

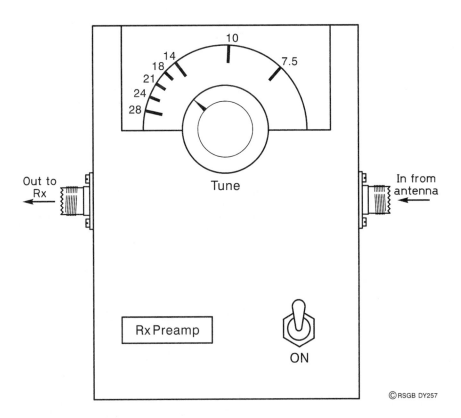

Figure 2 HF preselector, front panel layout

Testing

Don't fit your preselector yet. Tune your radio to a broadcast station, preferably a fairly weak one. Disconnect the aerial and fit your preselector between the aerial and receiver. Switch it on and rotate the tuning knob slowly. You should find a position where your original station is received more clearly than before. If it doesn't work at all, recheck your wiring. Is there a positive voltage on the drain of the FET? If not, work back towards the positive battery terminal. Is there a voltage at the junction of L2 and R2? Is there a voltage at the junction of L2 and R3? Is there a voltage at the junction of R3 and the battery lead? If there isn't a voltage at that point, then you have probably mounted your switch upside down, and it is off, not on! It's a common mistake.

Calibration

This is not obligatory, surprisingly enough. However, if having a frequency scale appeals to you, then using an RF signal generator (or using the services of a friend who has one) is the simplest solution. Feed in a weak modulated

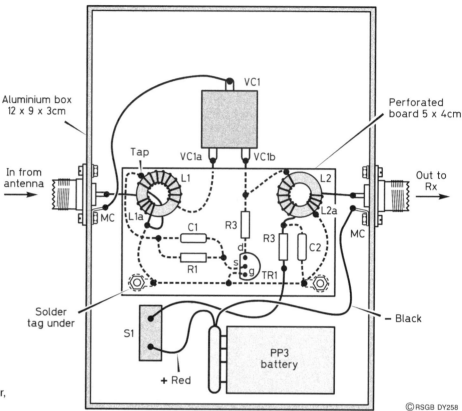

Figure 3 HF preselector, component layout

©RSGB DY258

signal from the generator to the preselector, and rotate VC1 until the signal is maximum. Mark this frequency on your dial. Repeat the process for the frequencies shown on the dial in Figure 2.

If you have a commercial transceiver, feed its output into a dummy load (such as the type described in *A Switched Dummy Load*, elsewhere in this book) using a distinctive modulating signal such as an idling RTTY signal. Set the receiver to the same frequency with the preselector out of circuit. Insert it into the aerial lead, and search for the signal with VC1 until it gives the maximum deflection on your S-meter, then mark the frequency on your dial.

Parts list

Resistors: all 0.25 W carbon film or better
R1	270 Ω
R2	15 Ω
R3	100 Ω

Capacitors
  C1, C2   100 nF (0.1 μF) ceramic
  VC1   250/250 pF polyvaricon

Inductors
  L1, L2   19 turns 26 SWG enamelled copper on T.50.6 toroid
                L1 has a tap 4 turns from ground end
                L1a and L2a – 3 turns wound over previous winding

Semiconductors
  TR1   MPF102 FET

Additional items
  Matrix board   To fit aluminium box (see Figure 3)
  Aluminium box   12 × 9 × 3 cm
  Battery and connector   PP3 9 V
  S1   SPST on/off switch
  Knob   As required
  SO239   Sockets – 2 required

75 An audible continuity tester

Introduction

This is not the only continuity tester in this book. This alone attests to their use, so you may well want to experiment with several designs, then come up with one of your own! The very simplest form of continuity tester is probably a battery and a bulb in series, with the circuit being closed by connecting it to a fuse or other object being tested for continuity. The bulb could be replaced by a buzzer to give an audible indication. The current taken by the buzzer could damage some components, however. An ohmmeter can also be used, and is very popular for the purpose, as it indicates whether the circuit is low, medium or high resistance. This project has the advantages of indicating whether there is no continuity, some resistance or good continuity, and making an audible sound, so that you don't have to move your eyes from the circuit while making the test.

The circuit

Using only one integrated circuit, four components and a battery, this is a particularly simple circuit, as **Figure 1** shows. IC1 is usually used to flash an LED from a 1.5 V source, and to have a low current consumption. By changing the component values, IC1 is made to oscillate at audio frequencies, and we can hear these through the loudspeaker, LS1. Low-impedance speakers can be used, but result in an increased current drain, so the higher the impedance, the better.

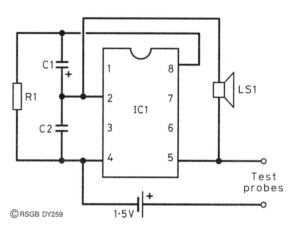

Figure 1 Continuity tester, circuit diagram

Construction

The layout of the circuit on Veroboard measuring 7 strips by 15 holes is shown in **Figure 2**. Start by making the four track cuts which will lie underneath the IC. Use a track cutter or a 3 mm (⅛ inch) twist drill rotated between thumb and forefinger for this. Solder the Veropins in place, followed by the wire link. Then solder in the IC holder, the resistor and the capacitors, making sure that the electrolytic capacitor, C1, is connected the correct way round.

Check, with the board against a bright light, that there are no shorted tracks, either by large blobs of solder or by copper swarf from the track-cutting process. Then insert IC1 into its holder the right way round. The probes can be ordinary connecting wire, the free ends being tinned with solder to prevent wire whiskers from touching components other than the one you are testing.

If you are happy that the circuit and the wiring appear to be correct, fit the battery into its clip. Nothing should happen until the probes are touched together, when you should hear a note from the loudspeaker. If nothing happens, all you can do is to recheck your circuit, as there is nothing else to go wrong!

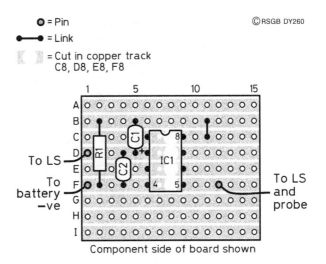

© RSGB DY260

O = Pin

●——● = Link

= Cut in copper track
C8, D8, E8, F8

To LS

To battery −ve

To LS and probe

Component side of board shown

Figure 2 Continuity tester, component layout

Any box can be used; there is no justification for a metal case, unless you want to make use of something, which is to hand, such as a tobacco tin. Alternatively, any suitable plastic box will do.

Use

The tester will give different pitch notes for resistances of different values. The higher the resistance, the higher the pitch from the speaker. Try it with small inductors, and you will learn to recognise the different tones produced by the IC.

Safety notice

Using the continuity tester on components *in situ* is not advisable, as the results could be misleading. It can be dangerous to make measurements *in situ* on equipment, which is operating. If you must make such tests, always disconnect the equipment form its power source first.

Parts list

Resistor
  R1   1000 Ω (1 kΩ), 0.25 W carbon, 10% or better

Capacitors
  C1   10 μF electrolytic, 16 V
  C2   0.1 μF subminiature polyester or ceramic

267

Integrated circuits
   IC1   LM3909N

Additional items
          Veroboard 7 strips by 15 holes
   LS1   64 Ω miniature loudspeaker
          IC holder
          Veropins
          Connecting wire
          Test leads
          1.5 V AA alkaline battery

76 An experimental 70 cm rhombic aerial

Introduction

Most commonly used aerials can be classed as *resonant* or *standing-wave* aerials. There is another class known as *non-resonant* or *travelling-wave*. Resonant aerials, such as the dipole, are narrow-band; this occurs because resonance occurs only over a narrow band of frequencies. Travelling-wave aerials, on the other hand, can operate over a wide band of frequencies.

The *rhombic* is an example of a non-resonant or travelling-wave aerial. It is often employed for fixed commercial and military short-wave radio links. Made with wire, it has a diamond shape when looking down on it from above. The four corners are supported on four masts. It is a very effective aerial, and has good gain, a quality which can be judged from the *polar diagram* shown in **Figure 1**. This has been obtained from a computer program, and so is the perfect shape for a rhombic aerial.

Theory (but only a little)

A polar diagram shows graphically the ability of an aerial to radiate (or receive) more effectively in one direction at the expensive of the radiation in other directions. Figure 1 shows the polar diagrams of two aerials, a simple 70 cm dipole and the rhombic described in this project. The dipole has the

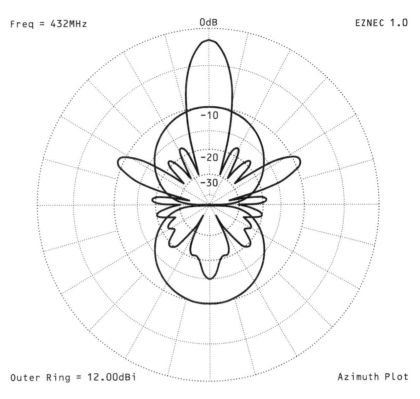

Freq = 432MHz   0dB   EZNEC 1.0

Outer Ring = 12.00dBi   Azimuth Plot

© RSGB DY262

Figure 1 The horizontal polar diagrams of the small rhombic antenna and a dipole compared

well-known figure-of-eight shape, showing a symmetrical angular behaviour about the direction of the aerial and about a direction perpendicular to the aerial. That of the rhombic, on the other hand, is quite irregular by comparison, but is still symmetrical about one axis only, not two. It's this asymmetry that gives the rhombic its gain, by virtue of its *front-to-back ratio*. This is the ratio of the power radiated forwards to that being radiated backwards. Notice the large *lobe* (lump) at the top of the polar diagram; this is the direction in which the aerial transmits best. The lobe in the opposite direction has been reduced significantly, allowing more power to be directed forwards, not backwards

Problems

The rhombic is not found in every amateur's back garden, despite its attractions. To work best each edge of the diamond shape should be about two wavelengths long. Hence, a rhombic for the 20 metre band could be about 80 m from tip to tip! Another disadvantage is that, because of its size, it cannot be rotated.

It will operate well over a range of frequencies. An HF rhombic could work on the 7, 10, 14, 18, 21 and 28 MHz bands. One for the lower VHF frequencies could operate on the 50, 70 and 144 MHz bands.

Its large size is less of a problem at UHF. A portable rhombic can be made for 70 cm which can be used with a hand-held transceiver. The aerial design to be described here will give a gain of up to 9 dB relative to a dipole. This is written as 9 dBd, the second 'd' meaning 'relative to a dipole'. This is equivalent to improving your signal by 1.5 S-points at the receiver or (and you may be surprised by this) by fitting a linear amplifier to your transceiver that would take 5 W input and produce 40 W output to a normal dipole! Consider the relative costs of the two approaches to producing the same received signal. You will also receive everyone else's signals 1.5 S-points better than before!

Unlike the popular Yagi aerial, this design has no critical dimensions and can be folded up for transport by car or bicycle. It has a fairly high input (or feed) impedance, being fed usually by balanced twin feeder. To feed it with standard 50 Ω coaxial cable, a matching transformer in the form of a *balun* is required. A balun will convert the *bal*anced (symmetrical) aerial impedance to the *un*balanced and lower impedance of the coaxial cable. The details of how to build the balun, which in this case is a half-wave transformer, are included in the constructional details.

Construction

The aerial frame is made up of 1 cm × 2 cm strips of wood fixed to a plywood centre using 30 mm long M4 bolts, as shown in **Figure 2a**. The outer bolts fixing the front and side supports can be removed for folding prior to transportation, as shown in **Figure 2b**.

The wires are fixed to the front and rear supports using screw connectors, sometimes called 'chocolate block' connectors. Detail X and Y of Figure 2 show how this is done. The side supports have holes in the ends through which the wire is threaded.

The aerial must be mounted in the horizontal plane using a small shelf bracket attached to the centre plate, using the same bolts that hold the rear support. The other half of the bracket may be mounted to a vertical mast using screws or jubilee clips (hose clips).

The balun

Cut a 23 cm length of coaxial cable; this is to be our half-wave transformer. Cut and remove 2 cm from each end of the sheath. Make the braid into a pigtail at each end. Cut and remove 1 cm of the inner insulator from each

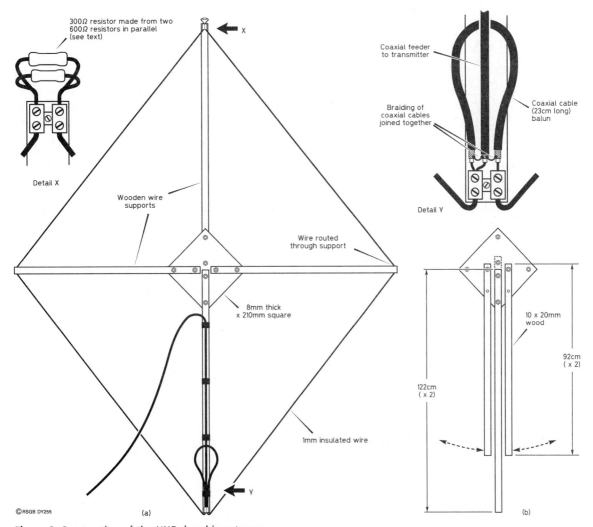

300Ω resistor made from two 600Ω resistors in parallel (see text)

Detail X

Wooden wire supports

Wire routed through support

8mm thick x 210mm square

1mm insulated wire

©RSGB DY255

(a)

Coaxial feeder to transmitter

Braiding of coaxial cables joined together

Coaxial cable (23cm long) balun

Detail Y

10 x 20mm wood

92cm (x 2)

122cm (x 2)

(b)

Figure 2 Construction of the UHF rhombic antenna

end, leaving 1 cm of the centre conductor exposed. Fasten this piece of cable to the chocolate block connector together with the coaxial feeder cable as shown in Detail Y of Figure 2. The braids of all three prepared ends are soldered together, but are not connected to anything else. The inner conductor of the feeder from the transceiver is soldered to one end of the centre conductor of the 23 cm piece, and is connected to one end of the rhombic by one side of the chocolate block. The other end of the half-wave transformer also goes to the chocolate block, where it is connected to the other side of the rhombic loop.

The opposite corner of the rhombic, as shown in Detail X of Figure 2, shows that the loop is broken at the chocolate block, and is 'terminated' by two 600 Ω resistors in parallel. This makes the aerial a broad-band travelling-wave device, and gives it its directivity and gain. There is a rule of thumb, which includes a safety factor, which says that the terminating resistor must be able to absorb one-half of the maximum transmitter output power. So, if you use two 2 W 600 Ω carbon resistors (*not* wire-wound resistors) in parallel, you can use a transmitter with an output of 8 W, which is more than adequate for 70 cm hand-helds.

Using the rhombic

Fit the aerial to a pole or mast in the horizontal plane, if you intend to use CW or SSB, but in the vertical plane if you want to concentrate on FM work. In the latter case, a wood or fibreglass pole is mandatory. Tune to a local repeater, whose signal strength you know. Rotate the aerial to face the repeater, and you should see that the signal strength is greatly improved! Verify the directional properties of the aerial by rotating it and observing the changes in signal strength on your S-meter.

77 Water level alarm

Introduction

This is a simple device which, when you have built it, you begin to wonder how you ever managed without it! It will sound a buzzer when the level of water reaches a particular point, which you can set and vary at will.

Detecting water

Water is not too difficult to detect electrically, because it is a conductor of electricity. Not a good one, but sufficiently so to carry enough current to control a warning circuit.

The circuit is shown in **Figure 1**. If you imagine a resistor placed between the ends marked 'water detector', the circuit is then recognisable as a transistor used as an electronic switch. Whether the switch is off or on depends on the value of this resistor. If the resistance is low, the transistor switches on and the buzzer sounds. If the resistance is high, the transistor switches off and the buzzer stops.

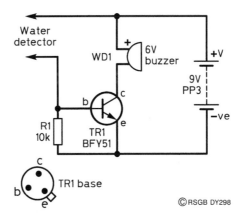

Figure 1 Circuit diagram of the water level detector

Construction

The unit is made in two separate sections – the detector board (**Figure 2**) and the main board (**Figure 3**). The main board is plain perforated board (without copper strips). The components are pushed through holes in the main board and connected together on the underside. The battery is held on with sticky tape. No on/off switch is used, as the circuit consumes virtually no current when it is not sounding the buzzer.

Unlike some buzzers, this type is polarised, and the red lead must go to the supply as shown in Figs 1 and 3.

The detector board is a piece of standard matrix board – the type with copper strips along one side. It measures 9 strips by 15 holes, and is wired in such a way that alternate strips are connected together, as Figure 2 shows. This gives a greater surface area for the water to touch, thus increasing the sensitivity of detection.

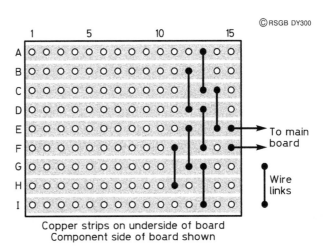

Copper strips on underside of board
Component side of board shown

Figure 2 Layout of the main board

273

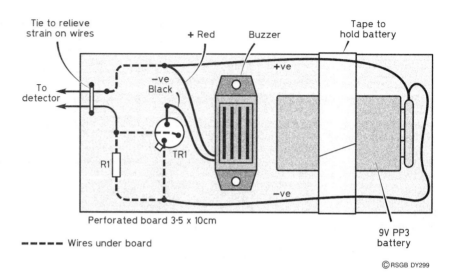

Figure 3 Water detector constructed from stripboard

Tie to relieve strain on wires

To detector

−ve Black

R1

TR1

+ Red

Buzzer

+ve

−ve

Tape to hold battery

9V PP3 battery

Perforated board 3·5 x 10cm

- - - - Wires under board

©RSGB DY299

The two boards are connected by ordinary twisted pair, enabling the detector to hang over the edge of the water container, with the buzzer circuit safely out of the way. When positioning the detector board, the left-hand edge (as shown in Figure 2) should face downwards, with the connecting-wire edge facing upwards. When the water level rises, the water causes current to flow between adjacent strips, and the higher the water, the lower the resistance between them. Eventually, the resistance will be low enough to turn on the transistor and sound the alarm.

Parts list

Resistor
  R1   10 000 Ω (10 kΩ), 0.25 W

Semiconductor
  TR1   BFY51

Additional items
  WD1   6 V miniature buzzer (polarised)
            Battery PP3
            Battery connector
            Strip board 9 strips by 15 holes
            Perforated board 3.5 × 10 cm

78 A delta loop for 20 metres

Introduction

HF aerials take up a lot of room when they are straight. Space can be saved by bending them and, at the same time, giving them properties which are quite different from their original linear forms. The delta loops shown in the accompanying diagrams have the shape of the Greek upper-case letter delta (Δ) upside down. It can be upright, except that it is more practical to have the feed point at the bottom than at the top.

Putting the loop together

This is the design for a loop to be used on the 20 metre band. It is a very popular band and carries most of the amateur radio DX traffic. The aerial is light in weight and operates best if the top section is at least 30 ft above the ground, with the feeder point being about 6 ft above ground. All you need is some good wire, some polypropylene rope, some insulators and a method of matching the output of this 75 Ω balanced aerial to the 50 Ω unbalanced coaxial input of your transceiver or receiver.

Each of the three sides must be 7.2 m long. Using tent pegs or six-inch nails, mark out the three corners on the lawn, making sure that each pair of pegs or nails is exactly 7.2 m apart. Put the wire around the nails, together with two insulators on the wire for the two top corners. These insulators can be of the 'dog bone' variety, or of the home-made type, using a piece of flat plastic with holes in each end. The use of these is shown in **Figure 1**. Located at the positions of two of the tent pegs or nails, they must be secured at those points by any convenient means.

Now attach a support rope to the free end of each of the insulators, ready for suspension, as shown in the diagrams, from any convenient supports (house, tower, tree, etc.). The top section of the loop could be taped to a continuous piece of rope between the two supports. You might like to try this if you think it is easier. Then, when you find some dog-bone insulators, you can change the design and see if there is any noticeable change of performance.

Connecting to the radio

If you are the proud possessor of an aerial tuning unit (ATU) with a balanced output, all you will need is a length of twin cable soldered to the

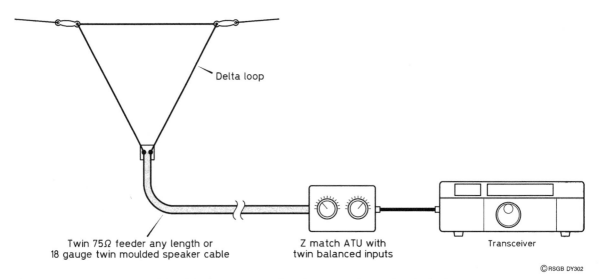

Twin 75Ω feeder any length or
18 gauge twin moulded speaker cable

Z match ATU with
twin balanced inputs

Transceiver

©RSGB DY302

Figure 1 Multiband version of the delta loop connected to the ASTU using 75 Ω twin feeder

ends of your delta loop and connected to the balanced output terminals on the ATU, as Figure 1 shows.

Another way is to buy a ferrite-cored 1:1 balun, and use it as indicated in **Figure 2**. This produces an aerial which will operate over the whole range of amateur and commercial short-wave frequencies, when used with an ATU.

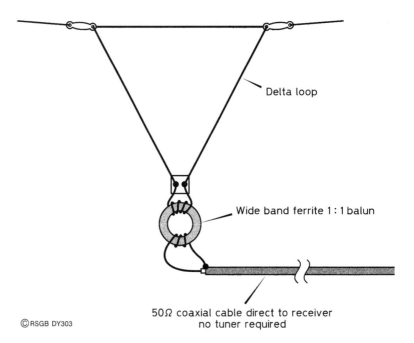

Delta loop

Wide band ferrite 1 : 1 balun

Figure 2 Multiband version of the delta loop using 75 Ω coaxial cable. This arrangement requires an ASTU if used with a multiband antenna

©RSGB DY303

50Ω coaxial cable direct to receiver
no tuner required

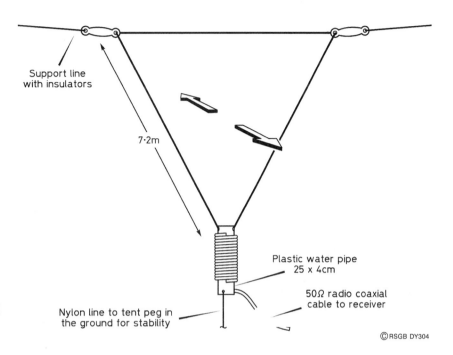

Support line
with insulators

7·2m

Plastic water pipe
25 x 4cm

50Ω radio coaxial
cable to receiver

Nylon line to tent peg in
the ground for stability

©RSGB DY304

Figure 3 Single band version of the loop antenna using a coaxial balun. The direction of maximum signal strength is indicated by the arrows

However, since this aerial is designed to be a single-band type, there is a simpler answer to the problem of matching the aerial to your 50 Ω coaxial cable – **Figure 3** shows this. Make a tuned balun with a length of plastic water pipe, 25 cm long and 4 cm diameter and a piece of good-quality TV aerial cable. TV aerial cable has a characteristic impedance of 75 Ω, compared with the 50 Ω of the coaxial feeder that we usually use with amateur radio equipment. Use the type with a brown sheath and a closely knit earth braid, not the type having an earth foil inside the sheath.

The length we need for the balun is 3.8 m, but always allow between 3 and 4 cm extra for preparing the ends. Drill two small holes diametrically opposed in the top of the tube; these will be used to anchor the two ends of the delta loop, as shown in Figure 3. Drill another single hole in the bottom end of the tube, which will be used to anchor a nylon line going to the ground to add stability to the loop. Then, after drilling a pilot hole, drill a 5 or 6 mm hole near the top end of the tube, as shown in **Figure 4**. Prepare both ends of the coaxial cable, then feed one end into the tube, far enough for its ends to be soldered to the ends of the loop when the assembly reaches that stage.

Now, close-wrap the cable around the tube until only about 3 cm remain. Holding the cable tightly, drill another 6 mm hole beside the free end of the cable and feed it into this hole. At this point, feed the two ends of the loop into the top two holes and twist it back on itself. Figure 4 shows how this is done. Then, solder the ends of the coaxial balun to the ends of the loop.

277

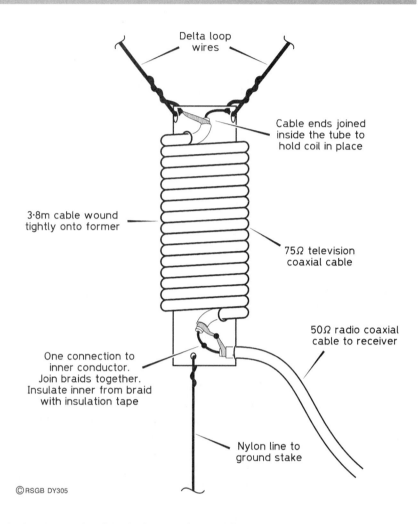

Delta loop
wires

Cable ends joined
inside the tube to
hold coil in place

3·8m cable wound
tightly onto former

75Ω television
coaxial cable

50Ω radio coaxial
cable to receiver

One connection to
inner conductor.
Join braids together.
Insulate inner from braid
with insulation tape

Nylon line to
ground stake

Figure 4 Construction of the
balun

©RSGB DY305

The bottom ends of the balun are then soldered to the 50 Ω coax which goes to your shack and to the transceiver.

Hoist the aerial into position carefully, being careful not to pull too hard on the support lines. Then, take the nylon line from the bottom of the balun to a peg in the ground. This adds stability to the aerial.

Using the delta loop

It is a directional aerial, as Figure 3 shows. It produces maximum power (and has maximum receive sensitivity) along a direction perpendicular to its own plane. Don't be too concerned with which direction it is pointing at first. Give it a try 'on the air' and see how it performs. Then you can contemplate how to point it in your favourite direction, to the USA, or Australia, for example.

Experimenting

You may want to enclose the balun in some sort of weatherproof container. Plastic ice cream containers are favourites for this sort of job. Seal all the holes where wires enter it with silicone sealant or self-amalgamating tape.

If you have used ordinary single-strand or multi-strand wire for your aerial, it will stretch over time under its own weight and that of the balun. Its operating frequency will fall slightly as a result. If you notice a significant difference, then dismantle it, remeasure and fix the sides and erect it again, perhaps facing a different direction. You can buy pre-stretched or hard-drawn wire for such purposes, if you feel that periodic tweaking of your aerial is a chore.

You can make a delta loop for different frequencies simply by scaling the lengths of wire for the loop and for the balun according to the design frequency. If your maths isn't quite up to this, enlist the help of someone well versed either in maths or aerial design!

79 A simple desk microphone

Introduction

A hand-held transceiver is usually the first type used by most novices. It is ideal when used out of doors, in the shack or mobile. However, using one in the shack with an outside aerial is not the best of solutions.

This article describes the design of a desk microphone which can be used either for transceivers using a common microphone and push-to-talk (PTT) line, or for those using a separate PTT line.

It is a simple design, using the barrels of two plastic pens, the lid of an old coffee jar, and the plastic sheath of a DIN plug.

Construction

The end caps and the ink tubes should be removed from two ball-point pens, leaving only the two plastic shells. One of them is used for the vertical microphone support, while the other is used to hold the microphone element.

The vertical support should have the pointed end shaped so that it holds the microphone support at a convenient angle, as **Figure 1** shows.

The base should be prepared next – see **Figure 2**. Four holes are required. Two, A and B, are small holes for the cable to be fed through. A third, C, is slightly larger and is used to take one of the plastic end caps. The fourth hole, D, is determined by the type of switch that you want to use for the PTT function.

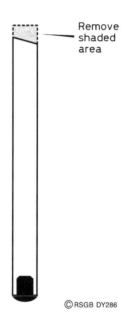

Figure 1 Vertical support made from ball-point pen shell

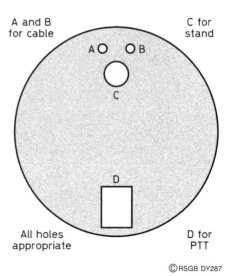

Figure 2 The microphone stand base

Some epoxy resin such as Araldite Rapid is used to mount the separate parts. Mix only a small amount at a time, and glue each part of the base assembly and allow to harden before moving on. Be sure to read and follow the instructions for the use of the glue to prevent accidents.

Firstly, glue the end cap to the base, feeding it underneath and pushing it up through hole C. Then, the vertical pen body is placed over the protruding end cap and glued into place, making sure that the top (shaped) end is pointing in the right direction to hold the other plastic tube carrying the microphone. While the glue hardens, place the DIN plug sheath on the second pen body. Place and glue this pen body on the shaped top of the vertical support, making sure that the whole assembly is supported while the glue sets. What you should have now ought to resemble **Figure 3**, but without the microphone and wiring. Leave everything to set for 24 hours.

Any switch of the press-to-make variety can be used for your PTT switch, but choose one which does not apply too much force to operate! If you find the whole assembly is too light and slides over your table, some plasticine can be pushed into the base when all the wiring has been completed.

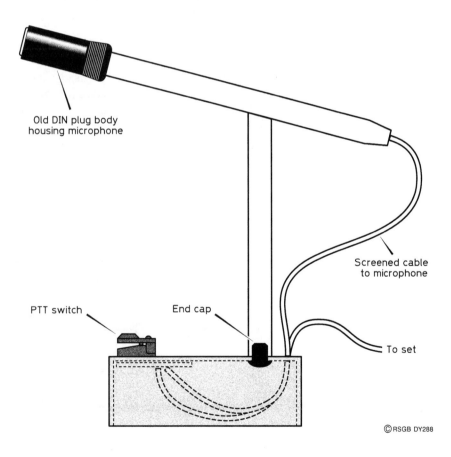

Old DIN plug body housing microphone

Screened cable to microphone

PTT switch

End cap

To set

Figure 3 Complete microphone stand assembly

©RSGB DY288

The circuit

A small electret condenser microphone is used, powered by a DC supply. Most hand-held radios use a single screened lead for both the PTT line and the microphone audio lead. In such cases, the circuit of **Figure 4** will operate the PTT function. Note that the PTT switch is in series with the audio lead from the microphone. If you find that the PTT switch does not switch your radio into transmit, reduce R1 from its normal value of $33\,k\Omega$ to $27\,k\Omega$. You should find that the original value works with most hand-helds.

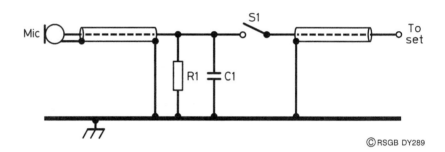

Figure 4 Wiring diagram for hand-held radios

©RSGB DY289

For radios with a separate PTT lead, the circuit of **Figure 5** is used. However, a power supply is needed for the electret microphone. This can be a PP3 battery, or can (in most cases) be derived from the microphone socket on the transceiver. You will need to consult the makers' handbook for this information. Figure 5 shows the PTT switch wired in a 'ground-to-transmit' configuration, which is correct for most base-station transceivers. If you're in any doubt about what your radio needs, consult an experienced friend.

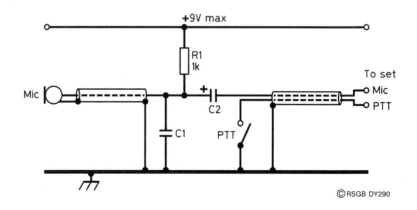

Figure 5 Wiring diagram for radios which use separate PTT lines

©RSGB DY290

The last connections

The microphone element should be connected with screened cable. Tape (masking or insulating) should be wound around the electret insert until it is a snug fit inside the DIN plug sheath. Feed the screened cable through the barrel of the pen until it emerges from the far end. Poke it through hole B and make the connections under the coffee jar lid. The components can all be mounted on the PTT switch or on a small piece of Veroboard mounted under the top surface. The output lead emerges through hole A and is of sufficient length to reach your transceiver. A suitable plug needs to be fitted to it.

Parts list

Resistors: all 0.25 W carbon, 5% tolerance
R1 (Figure 4)	33 000 Ω (33 kΩ)
R1 (Figure 5)	1000 Ω (1 kΩ)

Capacitors
C1 (Figure 4 and 5)	0.1 microfarads (0.1 μF) disc
C2	10 microfarads (10 μF) electrolytic

Additional items
S1	Push-to-make
Mic	Electret microphone (Maplin type FS34W)
Screened cable	As required
Plug	As required by your radio
Veroboard	If required
PP3 battery and clip	If required
Two old plastic ball-point pens	
Lid of coffee jar	
Plastic sheath from DIN plug	
Araldite or similar epoxy resin glue	

80 Morse oscillator

Introduction

This is not the simplest Morse oscillator to build, but it differs from the simple circuits in that it produces a pure note, not a coarse, rasping sound. People who have practised Morse using a non-sinusoidal oscillator sometimes find that they have trouble copying Morse code with a pure tone. As the pure tone is the correct way to receive Morse code, it is important that you should learn to listen to the code from a pure oscillator – so here's one!

The twin-T

There are many oscillator circuits, and there are many variations of the twin-T oscillator that we are going to use. **Figure 1** shows one version of a very useful circuit. All oscillators must have positive feedback in order to work. The feedback determines the frequency of the note produced by the oscillator.

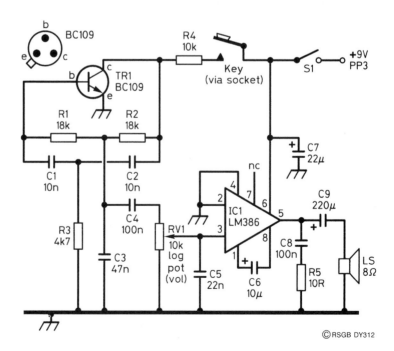

Figure 1 Circuit diagram of the sinewave oscillator and amplifier

© RSGB DY312

Here, the feedback circuit looks like two letter Ts. If you look at Figure 1, one T is formed by R1, R2 and C3, the other by C1, C2 and R3 – hence the name 'Twin-T'. Notice that the two Ts are connected in parallel between the collector and base of TR1, so any signal appearing at the collector is fed into both Ts. What emerges is then fed back into the base, producing in turn a signal at the collector. And so it goes on, producing a sine wave output.

The oscillator output is fed into an integrated circuit amplifier for output via a small loudspeaker.

Putting it together

The prototype was constructed on plain matrix board (the type *without* the copper strips), as shown in **Figure 2**. The components have their leads pushed through the holes in the board, and connections are made underneath.

Build the amplifier circuit first, using a socket for IC1. Connect the 9 V supply and touch pin 3 with an ordinary piece of wire. If a buzz is heard from the speaker, all should be well. If not, check your circuit and make changes until it does.

Build the oscillator circuit, and connect its output to the volume control VR1 via C4. Set VR1 half-way along its travel and switch on. A note should be heard from the speaker when the Morse key is depressed. The component values making up the twin-T determine the frequency of the note. Try varying them if you think your note is too high or too low. Whatever changes you make, either to the resistors or the capacitors, always ensure that R1 = R2 and that C1 = C2.

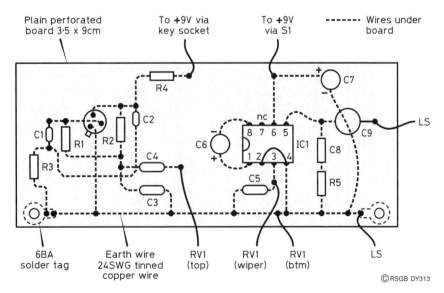

Figure 2 Component layout and interconnection diagram

©RSGB DY313

Parts list

Resistors: all 0.25 W carbon film
R1, R2	18 000 ohms (18 kΩ)
R3	4700 ohms (4.7 kΩ)
R4	10 000 ohms (10 kΩ)
R5	10 ohms (10 Ω)

Capacitors
C1, C2	10 nF (0.01 μF)
C3, C8	47 nF (0.047 μF)
C4	100 nF (0.1 μF)
C5	22 nF (0.022 μF)
C6	10 μF electrolytic, 16 V WKG
C7	22 μF electrolytic, 16 V WKG
C9	220 μF electrolytic, 16 V WKG

Semiconductor
TR1	BC109

Integrated circuit
IC1	LM386

Additional items
VR1	10 kΩ log potentiometer with switch
LS	8 Ω loudspeaker

Matrix board 3.5 × 9 cm
Small jack socket for key input
Box
PP3 battery and clip

81 A simple 6 m beam

Introduction

The attraction of building your own aerials is an abiding feature of our hobby. You can *buy* almost any shape or size of aerial, but one you have made yourself can often work every bit as well as a commercial device costing ten times as much.

The design

This aerial, designed for use on the 6 m band, is essentially a two-element Yagi, with the elements bent in order to reduce the physical size. It is known

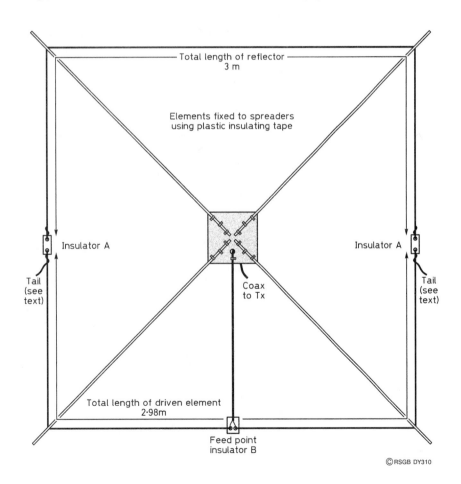

Total length of reflector
3 m

Elements fixed to spreaders
using plastic insulating tape

Insulator A

Insulator A

Tail
(see
text)

Coax
to Tx

Tail
(see
text)

Total length of driven element
2·98m

Feed point
insulator B

©RSGB DY310

Figure 1 Plan view of the complete 50 MHz VK2ABQ antenna

as the VK2ABQ beam, and was designed originally for the 20, 15 and 10 m bands, principally because of its space-saving qualities. It is made using a wooden frame and wire elements, and is ideal for portable operation.

Tools ready?

The beam is shown in **Figure 1**. The driven element is the one whose centre is fed by the coaxial cable, and lies between the two insulators marked A and the feedpoint at B. The reflector is also anchored at the points A, and lies over the upper half of the frame.

The wooden centrepiece is used to support the cross-pieces and to mount the aerial on the mast, using a common shelf bracket. The cross-pieces, known as *spreaders*, can be wooden canes or dowelling, and are mounted to the centrepiece using cable clips and adhesive. **Figure 2** shows how this is done.

If the aerial is to be a permanent installation, the spreaders should be weatherproofed using a good-quality exterior varnish. The wire elements are PVC covered and fed through holes in the spreaders.

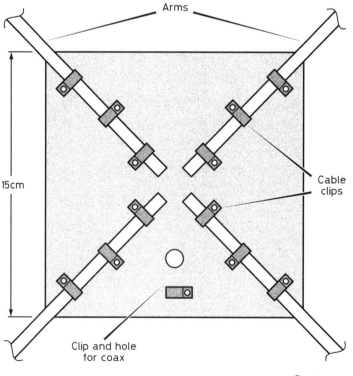

Figure 2 Centre support piece

©RSGB DY309

The end insulators are made of drilled perspex, and the wire passed through the two holes and twisted, as shown in **Figure 3**.

Note: if you have not drilled holes in perspex before, take care! The drill bit must be well-lubricated because it generates a lot of heat (enough to melt the perspex and jam the drill). Never turn the drill for more than a few seconds at a time, and moisten the bit in between. Start with a pilot hole and use bits of increasing size until the hole is the size you want.

Another perspex insulator is used to secure the feeder to the aerial, as shown in Figure 3. The feeder then passes directly to the centrepiece, where it is fastened with a cable clip and then passes down the mast.

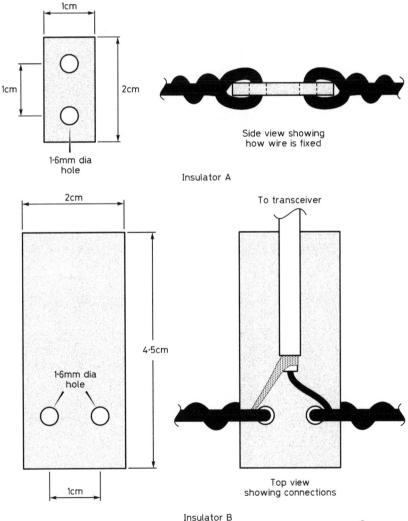

Figure 3 Details of insulators

© RSGB DY311

Adjustment

The driven element (A–B–A) in Figure 1 is fixed to the end insulators in such a way as to have 'pigtails' which are about 10 cm long. Using an SWR meter between the aerial and the transmitter, trim the pigtails equally at each end for minimum SWR in that portion of the 6 m band in which you plan to work. Make sure that the transmitter is off when you trim the ends, as high voltages can be present there. Always listen on the frequency before you transmit and, when you do, ask if the frequency is in use and identify yourself. Use as little power as possible.

The prototype had an SWR of 1.2 at 50.2 MHz and performed well. If you look at Figure 1, which is a view of the aerial from above, you will immediately see that when the aerial is horizontal, it radiates with horizontal polarisation in a direction from the top of the page to the bottom. A metal pole or mast can be used for horizontal polarisation, but if you intend to use the aerial for vertical polarisation, it is much better to use a wooden or fibreglass pole.

Portable use

If you plan to operate portable with this aerial, the only real modifications you need are to the centrepiece and how it supports the spreaders. Instead of using glue and cable clips, nuts and bolts through the spreaders and centrepiece would allow the spreaders to be 'hinged' closed for transport.

Materials

Centrepiece	Hardwood, $15 \times 15 \times 25$ mm
4 spreaders	110 cm long (cane or 6 mm dowel)
Wire	1.5 mm PVC-covered copper
50 Ω coaxial cable	RG58 or similar
Cable clips	6 mm (12 off)
Varnish	Polyurethane for waterproofing
Tape	Self-amalgamating, to waterproof all soldered joints
Insulators	See text

82 An integrated-circuit amplifier

Introduction

A simple audio-frequency amplifier is a very useful building-block in many more advanced electronic circuits. It is also a very useful piece of test equipment. To have one spare in the shack can be a life-saver at times.

Planning

Get into the habit of planning your project. How do you want it to look when it is finished. Do you want it in a box? Do you want it 'open-plan', with all the components on view?

The 'minimalist' approach to any project is simply a front panel and a baseboard, on to which all the components are fitted. In this case, the front panel would accommodate the loudspeaker, the volume control and the input and output jack sockets, and the baseboard would support the circuit board. Once you have decided these things, and know the size of the board, speaker and volume control, you can decided how big the panel and baseboard should be.

The amplifier

Instead of building the amplifier from discrete components (i.e. transistors), as was done in the project *An audio-frequency amplifier* (which you will find elsewhere in this book), we are going to use an integrated-circuit (IC) amplifier. External resistors and capacitors are still needed but, compared with the number of components *inside* the chip, these are very few indeed!

The IC we are going to use is the TBA820M. The circuit may look complicated, but with the use of a matrix circuit board it becomes quite simple. **Figure 1** shows the circuit diagram and **Figure 2** the layout on the board. The external connections to the PCB are shown in **Figure 3**.

To avoid having a separate switch to switch off the power supply when the amplifier is not being used, a volume control which incorporates a switch is used.

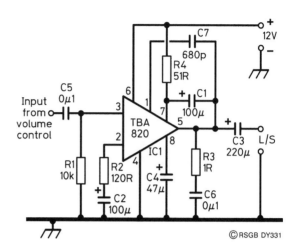

Figure 1 Audio amplifier, circuit diagram

©RSGB DY331

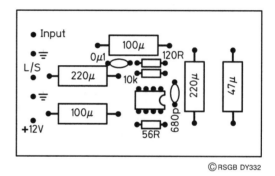

Figure 2 Audio amplifier, component layout

©RSGB DY332

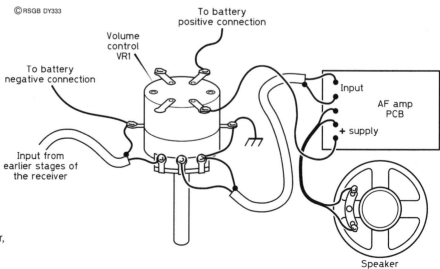

©RSGB DY333

Figure 3 Connecting the circuit board to the speaker, volume control/switch and battery

What power supply?

If the amplifier is to be used only for short periods, as a test instrument, for example, then it can be run from a PP3 battery, which can be mounted on the baseboard. If you intend to use it often, then a connection to an external power supply is preferable. For this option, you may want to consider fixing two terminals to the baseboard for this connection.

83 A novice ATU

Introduction

Having an aerial tuning unit (ATU) is always useful. It is used for adjusting the aerial impedance, as 'seen' by the transceiver, to be the same as that of the transceiver itself, usually $50\,\Omega$. This process is called *matching*, and ensures that the receiver and the power amplifier (PA) stage of the transmitter work efficiently.

This design is due to the late Doug DeMaw, W1FB, and uses readily available components. It will handle up to about 5 W, and operates over the frequency range 1.8 to 30 MHz.

Circuit evolution

The basic circuit of one type of ATU is shown in **Figure 1**. On transmit, the input signal at L1 is coupled into L2, which forms a resonant circuit with C1. The signal across the resonant circuit is fed to the aerial by C2. The combination of L2 and C1 helps to remove signal harmonics, because they are not at the resonant frequency and are shunted to earth. On receive, only those signals which are within the pass-band of L2 and C1 will pass into the receiver. This improves receiver performance by rejecting many out-of-band signals which the simpler receiver doesn't like.

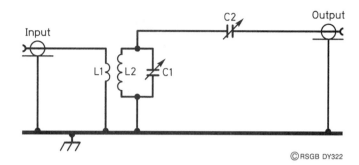

Figure 1 Basic circuit of the ATU

©RSGB DY322

For this design, the all-important resonant frequency is chosen to be in the 80 metre band. If we want to make the circuit operate on other bands, we need to change either L2 or C1. It is easy to change L2 by using a set of switched inductors (L3–L5) as shown attached to S1 in **Figure 2**. Forget about S2 and L6 for the moment.

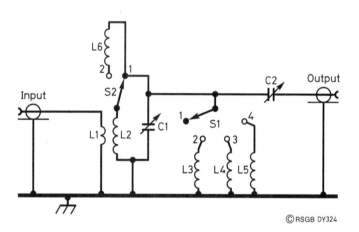

Figure 2 L6 added for 160 m

©RSGB DY324

When S1 is in position 1, the circuit reverts to that of Figure 1. S2 should be brought into circuit only when S1 is in position 1. Looking at the way the circuit is drawn, you can see that when S2 is in position 2, we have two coils in series, which is equivalent to adding more turns to L2. More turns means more inductance, which lowers the resonant frequency still more, and gives coverage of the 160 m band.

Rather than having to remember to flick S2 to the correct position *and* move S1 to position 1 when we want to operate on 160 m, both functions can be combined into one rotary switch with two wafers. This circuit is shown in **Figure 3**.

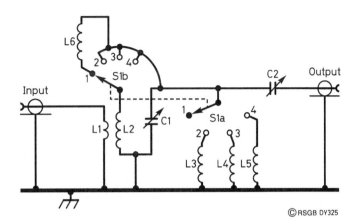

Figure 3 Two-pole switch for all bands

In all positions of S1a *except* position 1, the extra coil, L6, is shorted out. The two halves of the switch, S1a and S1b, move together, as the shaft is turned, so both halves are in position 1 at the same time, position 2 at the same time, and so on. The two switches, S1 and S2, are said to be *ganged*.

Construction

L1 is formed by winding four turns of 22 SWG enamelled copper wire over L2, as in **Figure 4**. L2 already exists on the purchased former. After scraping the enamel off the ends of the wire (with a sharp knife or sandpaper), one end of L1 must be soldered to one end of L2, as shown in Figure 4. The free end of L1 goes to the transceiver aerial socket.

All components except the capacitors are assembled on the switch. Note that the rotor of C1 is earthed, but neither side of C2 is earthed. This means that the metal shaft of C2 is not earthed, and touching it will detune the ATU. Using a plastic knob for C2 will minimise this effect. If you decide to use a metal box, take precautions to ensure that no part of C2 is in electrical contact with the box.

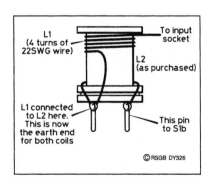

Figure 4 L1 is wound over L2 as shown

The capacitors are mounted using M2.5 screws. Make sure that the screws do not foul the vanes of the capacitor. If your screws are too long, a few washers between the box and the capacitor will solve the problem!

In use

The best indication of a good match is obtained with a standing-wave meter between the ATU and the transceiver. The controls are adjusted alternately to 'feel' your way to a better and better match.

For receive-only use, the same alternate adjustments are used, watching for the maximum signal strength on the S-meter or, for a very weak signal, making the signal from the loudspeaker as large as possible.

Parts list

Capacitors
  C1, C2   350 pF variable

Inductors
  L1   See text
  L2   27 µH
  L3   10 µH
  L4   2.2 µH
  L5   1 µH
  L6   65 µH

Additional items
  SO239 sockets (2 off)
  S1   2-pole 6-way rotary
  Box   as required
  Plastic knobs (2 off)
  Stick-on feet (4 off)
  M2.5 screws for capacitors
  Screws, nuts and washers for mounting the input and output
    sockets.

---

84 CW QRP transmitter for 80 metres

Introduction

This is a relatively simple transmitter design having an output of 1 W. The design is not new, having been described before in other amateur radio publications. The components are all available new and the total cost should not exceed £15.

The circuit

Like other simple transmitters (see *An 80 Metre Crystal-Controlled CW Transmitter* and *A Breadboard 80 m CW Transmitter* elsewhere in this book) this one is crystal controlled. This assures frequency stability, but limits the usefulness of the transmitter. The key to increased frequency coverage without a conventional Variable Frequency Oscillator (VFO) is the use of a low-cost 3.58 MHz *ceramic resonator*. The 'pulling' range of a 3.58 MHz ceramic resonator covers the UK novice 80 m sub-band and some of the CW segment below 3.525 MHz.

A ceramic resonator is like a crystal, but not quite as stable in frequency. Its main advantage is its large pulling range.

The block diagram is shown in **Figure 1**. It is very similar to a crystal-controlled transmitter, and includes an oscillator, buffer and final amplifier.

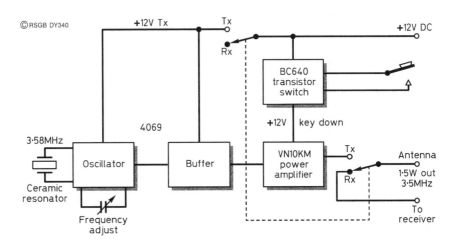

Figure 1 Block diagram of the ceramic resonator controlled 80 m CW transmitter

This amplifier is keyed, the oscillator remaining running all the time. This improves frequency stability because the oscillator is not being continuously stopped and started by the key. It is switched off while receiving, though, to avoid interference with the received signal. Transmit/receive switching is accomplished by a panel-mounted switch controlling both the aerial, oscillator and buffer switching.

Figure 2 shows the transmitter circuit diagram. An unusual aspect of this transmitter is the use of a digital CMOS integrated circuit (IC) type 4069 for the buffer and oscillator stages. The IC houses six inverters, four of which are used in the circuit. One is used as the oscillator, two are used for the buffer stage, and the fourth provides an output for a direct-conversion receiver, should one be added at a later date.

The frequency of the oscillator is changed by varying the capacitance in the ceramic resonator circuit. This is provided by VC1.

The power amplifier (PA) is a small MOSFET (metal oxide semiconductor field-effect transistor), TR1. This is capable of providing an output power of 2 W but, in this circuit, it is run conservatively to give 1.5 W. The output can be varied by changing the resistance (R5 + R6) in the gate circuit. Attempts

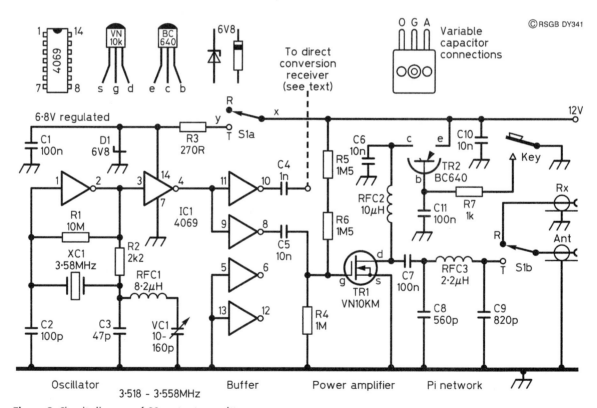

Figure 2 Circuit diagram of 80 metre transmitter

to raise the output power by decreasing the values of these resistors may result in immediate MOSFET failure.

A pi-network (C8, RFC, C9) provides impedance matching to 50 Ω, together with harmonic suppression. Like all inductors in this transmitter, the pi-network inductor is a pre-wound RF choke. A pi-network is so called because the components are arranged in the shape of the Greek letter pi (π).

Keying is carried out by a pnp transistor switch, TR2. Closing the key earths the base and supplies 12 V to the collector of TR2 and to the drain of the MOSFET, TR1, allowing the PA to operate.

Construction

You must house your transmitter in a metal box, to avoid hand-capacity effects and the radiation of spurious frequencies. Size is not important, provided it is large enough to accommodate the transmitter without cramping the components. You may want to allow space for future additions such as a direct-conversion receiver, break-in keying, sidetone or a small power amplifier. A good size is $5 \times 15 \times 15$ cm. You can make your own box, buy it, or even use a biscuit tin!

Front and rear panel connectors can be fitted first. The choice of these is a personal matter, but a good working choice would be:

(a) Power socket – 2.1 mm panel socket – centre pin positive.
(b) Key socket – ¼ inch jack socket.
(c) Aerial and receiver connectors – panel-mounting SO239 type.

Particular attention must be paid to the mounting of the variable capacitor, VC1. Make sure the hole for the shaft is amply big enough, and if you use screws to mount the capacitor on the front panel, then make sure they are not too long, otherwise they will touch the vanes of the capacitor! Mounting can be by means of glue, sparingly applied and kept well away from the shaft.

A board size of about 6×10 cm is adequate. Component layout on the board is suggested in **Figure 3**. The prototype used ordinary matrix board, which is preferable to stripboard for a design like this; stripboard has undue capacitance between adjacent strips. Component leads are fed through holes in the board and are connected underneath. Make sure that leads and connections are rigid because, if they can move, there is always the danger of short-circuiting, and capacitance changes.

To facilitate construction, servicing and testing, it is advisable to use Veropins for connections to the variable capacitor, transmit/receive switch,

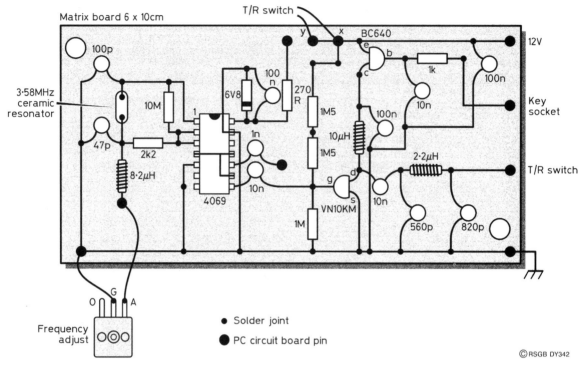

Figure 3 Component layout of the 80 metre transmitter. The transmit/receive switch is not mounted on the board and is not shown in this diagram

aerial and power sockets. Use screws and spacers to mount the circuit board to the box. Mounting the board horizontally assists troubleshooting.

Use a socket for the IC, and observe the CMOS handling precautions given in *An Electronic Die*, elsewhere in this book. When soldering the leads to the ceramic resonator, do it quickly – excessive heat damages the device. The earth lead running acoss the bottom of the board must be connected to the metal case by a short length of stout wire.

Testing

After carefully checking your wiring, both against the circuit diagram and the layout diagram, it is time to test your circuit. You will need a multimeter, an 80 m SSB receiver and a 50 Ω dummy load. A good design of dummy load can be found in the project *A Switched Dummy Load*, also in this book. An RF power meter and frequency counter will also be useful, although if your receiver has a digital frequency readout and S-meter, the latter two items are not really necessary. You will also need a 12 V 1 A power supply unit (PSU) to power the transmitter.

Switch the transmitter to receive and switch on the transmitter. No current should be consumed. Switch to transmit and check that pin 14 of IC1 is 6.8 V positive. With the dummy load connected to the aerial socket, press the key. The voltage on TR2 collector should now be 12 V, dropping to zero when the key is released.

Now check the operation of the oscillator. In transmit mode, you ought to be able to find a strong carrier signal with the receiver, even though the dummy load is connected. Adjusting the variable capacitor should change the frequency. At the lower end of the frequency range, you may find that the oscillator is unreliable in starting, because the circuit is attempting to pull the resonator too low in frequency. If this is the case, set the trimmer at the back of VC1 to minimum capacitance. If your version of VC1 has two trimmers, and you don't know which one to set, set them *both* to minimum capacitance. If there is still a problem, reduce the value of RFC1 to 6.8 or 4.7 μH.

In all probability, the unmodified circuit of Figure 2 will not require any of the changes outlined here. A coverage of 3.518 to 3.558 MHz should be possible, while preserving good frequency stability and reliable oscillation.

A signal probe (see *An RF Signal Probe*, elsewhere in this book) is useful for checking the operation of the oscillator and PA. Alternatively, an RF power meter or the receiver's S-meter can be used. With the PA running, the unit should draw between 200 and 300 mA. If TR1 becomes too hot to touch after a few seconds of transmitting, increase R5 or R6 to limit the transistor's heat dissipation. A small 6.3 V bulb connected across the aerial output is a simple way to check that the PA is working. An orange/white glow when the key is pressed is indicative of correct operation.

The final test is to monitor keying 'quality'. With your dummy load connected, press the key and listen to the note on the receiver's loudspeaker. Then operate the key, sending a string of dits, for example. What you hear should be free of chirps and clicks, as well as being stable in frequency. This test is sometimes better performed with no aerial connected to the receiver, thus preventing receiver overload and its associated plops. No problems should be encountered here.

Frequency tuning

A peculiarity of ceramic resonators is that, every now and again, their frequency changes abruptly by 100 Hz or so, then remains stable for some time. This is certainly noticeable in the received signal, but does not detract from the QSO and no characters are lost as a result. Try to keep the area around the ceramic resonator cool, to avoid temperature variations.

Parts list

Resistors: all 0.25 W, 5% tolerance

R1	10 megohms (MΩ)
R2	2200 ohms (2.2 kΩ)
R3	270 ohms (270 Ω)
R4	1 megohm (1 MΩ)
R5, R6	1.5 megohms (1.5 MΩ)
R7	1000 ohms (1 kΩ)

Capacitors

C1, C7	100 nF
C2	100 pF
C3	47 pF
C4	1 nF
C5, C6, C10	10 nF
C8	560 pF
C9	820 pF
VC1	10–160 pF variable

Inductors

RFC1	8.2 μH
RFC2	10 μH
RFC3	2.2 μH

Semiconductors

IC1	4069
TR1	VN10KM
TR2	BC640
D1	6V8 Zener

Additional items

Matrix board (see text)
14-pin DIL socket
Pointer knob
Sockets (see text)

85 An audio booster for your hand-held

Introduction

The audio output from many hand-held transceivers and receivers usually leaves much to be desired, so this little amplifier was designed to increase the output at minimal expense.

All that is needed is a separate amplifier and bigger loudspeaker. This is accomplished using a single integrated circuit (IC), a few components, and a loudspeaker from the junk box. This circuit will enable the output from your hand-held to be heard easily in a car.

The circuit

This is shown in **Figure 1**. It uses only those components necessary to operate the IC amplifier. VR1 is the preset volume control, and varies the signal coming from the 'External speaker' jack on the hand-held before feeding it into the IC for amplification. C1 blocks any constant voltage present on the input.

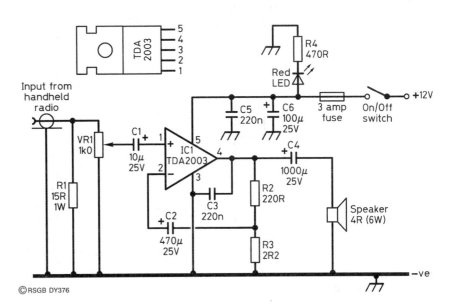

Figure 1 Circuit diagram of the audio amplifier. The power is derived from the cigarette lighter socket and the fuse is in-line with the lead

© RSGB DY376

303

The IC output comes from pin 4 and is fed via the electrolytic capacitor, C4, to the loudspeaker. The circuit is provided with an on/off switch, fuse and LED to indicate when the circuit is switched on.

Construction

The box is made of aluminium. This is necessary to help to dissipate some of the heat generated by the IC. Do not build the circuit inside a plastic box unless you take special precautions! The IC has a metal mounting tab with a hole, specifically designed to be mounted to a metal box or other metal heat sink. Apply plenty of heat sink compound between the tab and the box, tighten the nut and bolt, and then wipe off any excess compound. The box will get slightly warm in operation.

The size of the box is not specified. You may want to decide on this when you find a loudspeaker. Choose one which will be able to handle 6 W

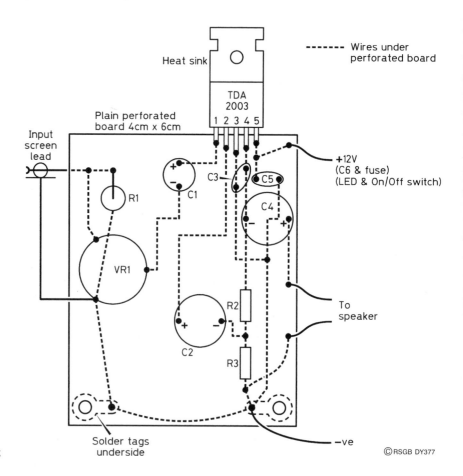

Figure 2 Layout of the components within the box

output. Drill all the holes in the box first. Holes for the speaker, input phono socket and the LED. The amplifier can be constructed on ordinary matrix board, which can be mounted inside the box with screws and spacers.

The layout is shown, for your guidance, in **Figure 2**. The components are mounted by pushing their leads through the holes in the board and making connections on the underside. The preset volume control, VR1, is set such that the hand-held's volume control is sufficient to control the final output over a good volume range. Use a screened lead from the 'External speaker' jack socket to the phono plug.

An external power supply is needed for this circuit. The normal dry battery which we usually use for small projects in this book will *not* work here, so you will need a proper mains power supply producing a stabilised 12 V. If you are going to use the unit principally in a car, then the cigar lighter socket can supply this voltage easily. Do make sure that the polarity is correct before you switch on!

When you plug the jack plug from your booster into the 'External speaker' socket on your hand-held, its internal speaker will be muted, so don't think that something dire has gone wrong! Adjust VR1 for a good volume range on your booster, when the volume control is turned on the hand-held.

Parts list

Resistors: all 0.25 W, 5% tolerance, unless otherwise stated

R1	15 ohms (15 Ω) 1 W
R2	220 ohms (220 Ω)
R3	2.2 ohms (2.2 Ω)
R4	470 ohms (470 Ω)
VR1	1000 ohms (1k Ω)

Capacitors

C1	10 μF 25 V
C2	470 μF 25 V
C3, C5	220 nF (0.22 μF) Mylar
C4	1000 μF 25 V
C6	100 μF 25 V

Semiconductor

LED	5 mm Red

Integrated circuit

IC1	TDA2003

Additional items
   Heat sink compound
   Nuts and bolts
   Loudspeaker 4 Ω 6 W
   3 A fuse
   On/off switch (SPST)
   Matrix board 4 × 6 cm
   Solder tags
   Plugs and screened cable for connecting lead
   Aluminium box

86 A grid dip oscillator

Introduction

When an inductor and a capacitor are connected, whether in series or parallel, they form a circuit with a natural (or *resonant*) frequency. The circuit stores energy, and this energy is being constantly shifted from the inductor to the capacitor and back again.

The dip oscillator is a simple instrument used to measure the resonant frequency of a tuned circuit without having to make any direct connection to the circuit. The circuit is more commonly known as the *grid dip oscillator* (GDO), from the days when the active device in the circuit was a valve. The FET or Field-Effect Transistor operates in a way which is very similar to that of the valve, so it is not quite a misnomer to call this instrument a grid dip oscillator, too.

The circuit

The GDO uses a calibrated, tunable FET oscillator in the circuit of **Figure 1**. It has a frequency range of 1.6 to 35.2 MHz in four ranges using a set of plug-in coils, shown in **Figure 2**. When the oscillator coil, L1, is placed near an external resonant circuit, some of its RF energy is coupled into the external circuit. A gain in energy of the external circuit must mean a loss of energy in the GDO circuit, resulting in a change of current through TR1, which is measured by the meter, M1.

The current through TR1 is of the order of 5 to 8 mA, but the change of current may be only a few microamps. To measure a very small

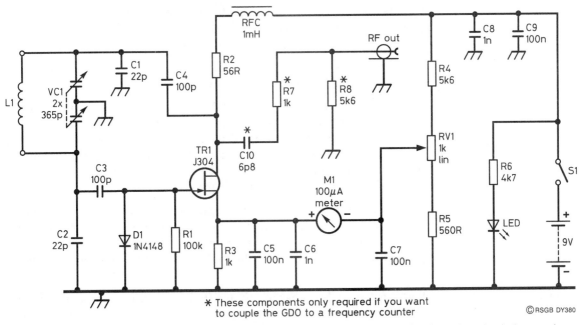

Figure 1 Circuit of an FET GDO. The coils are wound on DIN speater plugs, which provide both a plug-in base and a coil former

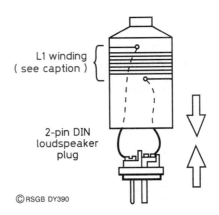

Figure 2 Details of coil construction:
Range 1: 1.6–4.0 MHz 55 turns of 30 SWG
Range 2: 3.3–7.9 MHz 27 turns of 30 SWG
Range 3: 6.3–4.0 MHz 14 turns of 26 SWG
Range 4: 11.9–4.0 MHz 7 turns of 24 SWG.

change superimposed on a much larger standing current, the method of *offset* can be used.

One connection to the meter goes to the source of the FET, while the other goes to a variable offset voltage set by VR1. M1 has a full-scale deflection (FSD) of 100 μA. If the current through TR1 changes, the voltage across R3 changes. When there is no resonance, the voltage at the wiper of VR1 is set to be very slightly greater than that across R3, and there is a 75% FSD meter deflection. When the voltage across R3 decreases very slightly, due to external circuit resonance, a significant 'dip' in the meter deflection is produced, hence the name of the instrument.

The circuit is not difficult to make on standard matrix board. Provided you can follow a circuit and translate it into a good component layout, then this project is probably only an evening's work.

The most important part of the GDO is the tuning capacitor and its associated frequency-calibrated dial. New, air-spaced tuning capacitors can cost you up to £20, so it is worth delving around in junk boxes, or scouring the tables at a local boot sale or rally. The tuning capacitor from an old transistor radio should be perfect. It may even have a slow-motion drive and a dial which can be remarked for the project.

Choose a coil plug and socket arrangement that is practical. Think about crystal holders or phono plugs and sockets. The prototype shown in Figure 1 and Figure 2 used 2-pin DIN plugs, with the coil wound on the outside of the plastic plug cover. Figure 2 shows the coil construction and the winding details.

If you use a variable capacitor, VC1, with a value different from that shown in the parts list, then the frequency ranges will be different. This does not matter, as it will be taken into account during calibration.

Position VC1 so that the dial will be easy to see and to operate, while locating the coil socket as close to it as possible. **Figure 3** shows the traditional layout of the GDO.

Calibration

Because the GDO also radiates a very small amount of energy, a general coverage receiver can be used to calibrate the dial. Don't try to aim for great accuracy and clutter the dial with marks and figures! If you include C10, R7 and R8 in your circuit, you can connect a frequency counter directly to the GDO and leave it in circuit all the time.

The GDO in use

Always try to place the external tuned circuit with its coil coaxial with the plug-in coil, as shown in Figs 1 and 3. If the coils are at right angles, the GDO may not produce any resonance. Set the offset control to give about 75% FSD and slowly tune L1 through its whole range. If no dip occurs, you may have the wrong coil plugged in. When you eventually find a dip, move the external coil further away until only a minute dip is seen. You may have to retune the dip meter as you do this, but it gives a much more accurate reading of frequency.

Remember that you cannot 'dip' a coil by itself – there must always be a capacitor present.

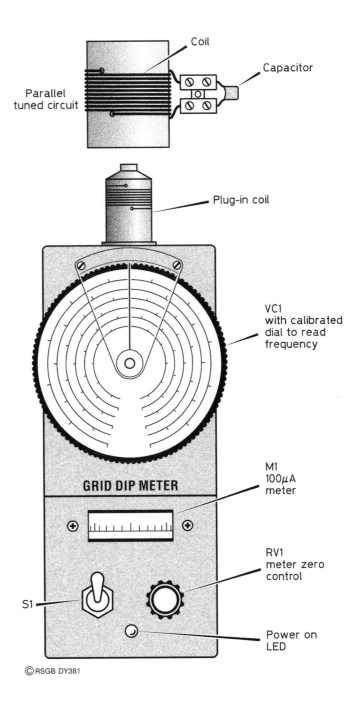

Coil

Capacitor

Parallel
tuned circuit

Plug-in coil

VC1
with calibrated
dial to read
frequency

GRID DIP METER

M1
100μA
meter

RV1
meter zero
control

S1

Power on
LED

©RSGB DY381

Figure 3 Layout of a typical
GDO. The dial, meter and
the location of the coil to
the circuit under test can all
be viewed at the same time.
It is shown measuring
resonance of a tuned circuit

Aerial resonance

Figure 4 shows how to check the resonance of a dipole aerial. Disconnect the coax feed at the aerial and place a short piece of wire, terminated with crocodile clips, across the centre insulator to short together the two ends of the aerial. By placing the GDO close to the shorting link, a dip should be seen on the meter while VC1 is turned. Alternatively, a loop in the element can be made around the coil, as in Figure 4, or the shorting link can be made long enough to loop over the coil. The latter method does not require tampering with the mounting and tensioning of the dipole wires.

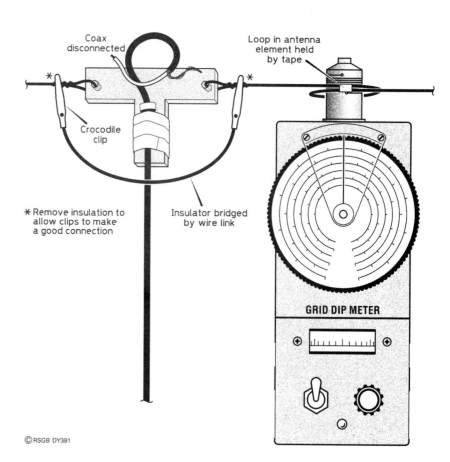

Figure 4 A wire antenna element can be looped into a single turn coil for increased coupling to the GDO

©RSGB DY391

Parts list

Resistors: all 0.25 W, 5% tolerance
R1	100 000 ohms (100 kΩ)
R2	56 ohms (56 Ω)
R3	1000 ohms (1 kΩ)
R4	5600 ohms (5.6 kΩ)
R5	560 ohms (560 Ω)
R6	4700 ohms (4.7 kΩ)
R7	1000 ohms (1 kΩ)
R8	5600 ohms (5.6 kΩ)
VR1	1000 (1 kΩ) linear

Capacitors
C1, C2	22 pF
C3, C4	100 pF
C5, C7, C9	100 nF (0.1 μF)
C6, C8	1 nF (1000 pF)
C10	6.8 pF
VC1	2 × 365 pF

Semiconductors
TR1	J304 or similar
D1	1N4148
LED	

Additional items
L1	See Figure 2
RFC	1 mH
S1	SPST
M1	100 μA

Source

Components are available from Maplin.

87 A CW transmitter for 160 to 20 metres

Introduction

This very small transmitter is designed to work on any band from top band (160 m) to 20 m, with an RF output of 1 W. It will work on higher frequencies but with a reduced output.

The circuit

The three-transistor circuit is shown in **Figure 1**. It comprises a crystal oscillator using a BC182 transistor which drives a 2N3866 power amplifier (PA) keyed by a ZTX750 PNP transistor. The oscillator and PA are coupled by a capacitor and resistor; this provides a very small amount of positive bias to the PA.

The oscillator can be used as a basic crystal oscillator but, by including a variable series capacitor as shown in Figure 1, the crystal frequency can be 'pulled' slightly, making the oscillator a *variable crystal oscillator* (VXO).

Construction

The PCB layout is shown in **Figure 2** and in the photograph. Although the prototype was built around a PCB, this circuit is equally amenable to

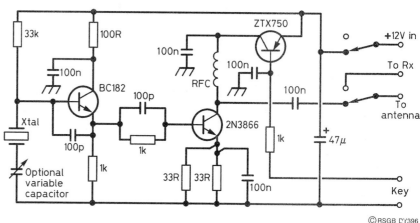

Figure 1 Transmitter circuit diagram

©RSGB CY396

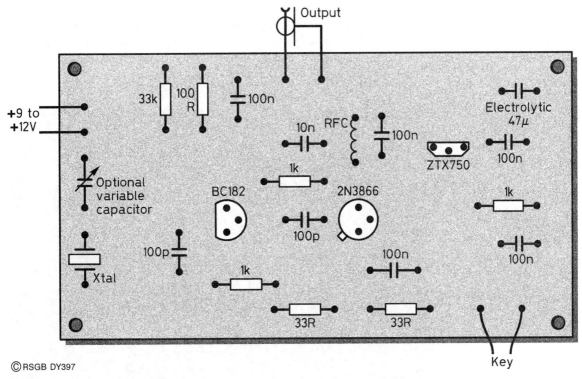

Output

33k · 100 R · 100n

+9 to
+12V

Electrolytic
47μ

10n · RFC · 100n

100n

ZTX750

Optional
variable
capacitor

1k

BC182 · 2N3866

1k

100p

Xtal · 100p

100p

1k

100n

100n

1k

33R · 33R

Key

© RSGB DY397

Figure 2 Component placing on the PCB. The external connections are also shown

construction on a matrix board. Populating the PCB is very simple, and you can expect to be able to do this in about one hour. The radio-frequency choke (RFC) is made by winding 10 turns of 33 SWG wire on a ferrite bead. The enamel coating of the wire is intended to vaporise when soldered into the board, thus obviating the need to remove the enamel manually with a knife or sandpaper. However, if you do have problems with the PA either not working or keying intermittently, it is suggested that you investigate the RFC connections immediately!

If you decide to make the VXO version, you will have to cut the track between the crystal and earth, and connect the variable capacitor (250 pF) across the break.

In use

After performing the usual checks on the accuracy of your circuit building and the wiring of the external components, it is time to connect a 12 V battery between the points shown in Figure 2. Do not switch on yet. An aerial needs to be connected to the output via an ATU and a crystal, of frequency matching that of the aerial, fitted. The variable capacitor should give you a tuning range from about 14.058 to 14.064 MHz. 'Netting' (the process of tuning your transmitter to the same frequency as that of a received station) is achieved simply, because the oscillator is always running, and the leakage of the signal (despite the fact that the PA is not powered) is sufficient to bring the two frequencies to zero beat.

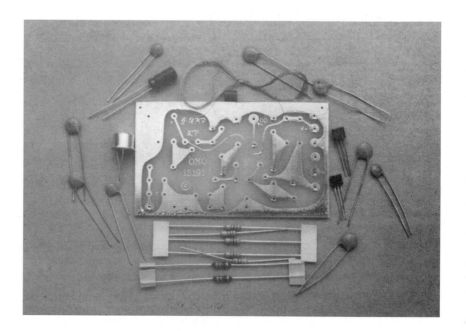

Warning

Note that the transmitter has no filtering; harmonics are not suppressed. It is strongly recommended that you use this transmitter in conjunction with the excellent low-pass filter described in the project *A 7-element low-pass filter for transmitters*, described elsewhere in this book.

88 Matching the end-fed random-wire aerial

Introduction

Many amateurs who do not have the space (or money) for a multi-band beam aerial, make use of the simplest possible alternative – the longest piece of wire that they can erect, with its end connected to the transceiver or receiver by an aerial tuning unit (ATU).

The length of the wire is not of major importance. Any length between 10 m and 80 m, with bends if necessary, will suffice. A good earth connection to the radio is just as important. Bends in the aerial wire can have some interesting effects on the directional properties of the aerial; V- or L-bends, or even a square shape are permitted. The only thing *not* permitted is to fold the wire back on itself in a tight hairpin bend!

Longish wire aerials

The term 'long wire' is usually used (incorrectly) to describe an end-fed aerial. How long is a piece of string? It depends what you mean by 'long'. In aerial parlance, it means 'long with respect to one wavelength'. Again, this depends on the band you are using. A long wire at 20 m is somewhat different from a long wire at 160 m. However, if you have sufficient real estate for a long wire on the 160 m band it must, by definition, be a long wire on all the other bands, too!

This should put you on your guard when analysing published data about feed-point impedance and the directional properties of a long wire aerial. Such theoretical data relate to a *real*, *ideal* long wire which is straight, horizontal, very high above perfect (conducting) ground, and not obstructed in any way. So your aerial doesn't quite match these criteria? Join the club!

Don't let this dampen your ardour when it comes to evaluating what the longish wire can do for you. The following should explain why.

Feed-point impedance

The impedance at the end of a longish-wire aerial can vary from a few tens of ohms to several thousand ohms, depending on the frequency in use and the wire length. It is also affected by factors such as bends, height above ground, proximity to buildings and wire diameter.

The actual value doesn't matter, provided we can make the aerial *appear* to have a 50 Ω impedance at the aerial socket of the transmitter. This process is what we call *impedance matching*, or simply *matching*. It maximises the power transfer from the transmitter to the aerial, and from the aerial to the receiver.

That is why an aerial tuning unit, or ATU, is almost (but not necessarily) obligatory.

The ATU

Many commercially produced HF receivers and transceivers have single 50 Ω coaxial sockets as their one and only means for connecting an external aerial. This means that an external aerial should have a 50 Ω feed impedance if it is to work efficiently, and it rules out most of the aerials being used by amateurs on the HF bands. Some means is necessary to change the aerial feed impedance to 'match' that of the transceiver. Such impedance-matching, or Z-matching (because Z is the symbol for impedance, just as L is the symbol for inductance) is the rôle of the ATU.

These can be bought and will accommodate either an end-fed or a coax-fed aerial. They can be bought ready for use or in kit form. Whether you want one for receiving only, or for use with a low-power (QRP) or high-power (QRO) transmitter, will determine what you need and how much you pay.

A simple single-band ATU

The simplest form of ATU is shown in **Figure 1**. It is simply a parallel LC (coil and capacitor) circuit, resonant at the chosen frequency, with taps on the coil for the aerial and the coaxial feed to the transceiver. If we assume the circuit is resonant, a high impedance exists at the top of the coil, and a low impedance at the bottom.

We said earlier that the end-fed longish wire presented an impedance which was high (or at least higher than $50\,\Omega$). This explains why the aerial is tapped to the coil near the top, where the impedance is high, and the $50\,\Omega$ coax is tapped near the bottom, where the impedance is low.

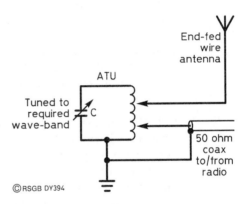

Figure 1 Parallel tuned circuit as single-band ATU for end-fed longish-wire antenna

Because the feed-point impedance of the aerial changes with frequency, so must the point at which the aerial is tapped to the coil to achieve impedance matching. The value of C must be changed also, to ensure that the circuit is resonant, and the $50\,\Omega$ tap will require tweaking also.

Setting up an ATU is quite simple. Make up an LC parallel-tuned circuit consisting of 50 turns of enamelled copper wire on an empty 35 mm film plastic container (or similar), tuned with a 500 pF variable capacitor. Make sure the enamel is removed from the ends of the wire before soldering.

Solder the inner wire of the coaxial cable from the radio to the first or second turn of the coil from its grounded end. Then solder the braid to the grounded end. Connect the aerial about one-third of the way down the coil from the top, removing the enamel at the connection point.

Adjust the variable capacitor for maximum noise or signal strength in the receiver. Then, try different tapping points from the aerial, to maximise the signal again. This matches the aerial impedance to that of the tuned circuit. Repeat the process with the coax tap, thus matching the impedance of the radio and the cable to that of the tuned circuit.

You will no doubt find that the tapping process on the coil was not easily accomplished, especially when the enamel must be removed at each tapping point without shorting adjacent turns together. It is therefore logical to produce a design where the taps have been prepared during the winding of the coil, and are selected with a rotary switch. To this end, the following multi-band design is described.

A multi-band ATU

The same type of coil design around a 35 mm film container is used (see **Figure 2**). The tapping points can be prepared in advance by a little judicious planning. Dismantle your original coil and measure the length of wire on it. It will be a little more than a calculation of $50 \times \pi D$ would suggest (where D is the diameter of the film container), due to the lack of tension in the coil and the wire diameter itself. You will need aerial tapping points at turns 1, 2, 3, 5, 10, 15, 20, 25, 30, 35, 40 and 45, corresponding to all the bands from 28 MHz to 1.8 MHz. The coaxial cable tap is fixed at turn 2 (an acceptable compromise). All the turns are counted from the earthy end.

Cut another piece of enamelled copper the same length as you used originally. Then, with the aid of an ordinary calculator, work out the positions of the points where the enamel must be removed for the taps. For example, turn 15 will have to be made $\frac{15}{50}$ of the way along the wire, turn 20 tap made at $\frac{20}{50}$ of the way along. So, if the length of wire is, say 4.7 metres, the two taps in question will be made at $\frac{15}{50} \times 4.7 = 1.41$ m and $\frac{20}{50} \times 4.7 = 1.88$ m from one end. This must be repeated for each of the tap positions, and the enamel removed ready for the wire to be soldered to it. With a

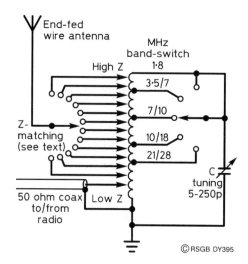

Figure 2 Simple multi-band ATU for end-fed longish-wire antennas

soldering iron, tin each tap point while the copper is shiny, thus ensuring a good, low-resistance connection. Do not solder on the taps yet.

The coil can now be wound as before. The taps can be soldered on, taking the lead from each one to the wafer of a single-pole 12-way rotary switch, the pole being connected to the aerial.

The tuning of the ATU is carried out by the same 250 pF capacitor, with a single-pole 5-way rotary switch used to select the band. Its tapping points will need to be chosen manually, using the method described earlier. Don't attempt to make new tapping points on the coil for this – use the taps available on the wafer of the other rotary switch, and find which is optimum for each band.

Notes

- For the aerial, use PVC-covered stranded tinned copper, of size 16/0.2 mm or 24/0.2 mm.
- Make the wire as long as possible, but anything over 10 m should be OK.
- Keep the wire as high as you can, in the clear and away from obstructions.
- Don't worry about bends, but don't use hairpin bends.
- Use a good insulator to attach your aerial – anything plastic will do.
- Anchor the wire near the point of entry to the building, but use a U-bend to prevent ingress of water.
- The wire can be brought in through the corner of a window, the PVC acting as an insulator. If you must drill a hole in the brickwork, make sure it slopes upwards from outside, so that water is deterred from entering.
- Use a good RF earth (as opposed to an electrical earth) such as half a dozen bare copper wires buried under the lawn in a fan shape. They should be joined together at the point of the fan and strapped to the earth connector of your ATU.